T0240058

Integrated Treatment Technology of Rural Domestic Sewage

Wensheng Li · Yungui Li · Jianmin Zhang ·
Fengyu Wang · Bin Wang

Integrated Treatment Technology of Rural Domestic Sewage

Ten Cases of Integrated Sewage Treatment
in Rural Area of China

Wensheng Li
Yunnan Hexu Environmental Technology
Co., Ltd.
Shenzhen, China

Jianmin Zhang
Yunnan Hexu Environmental Technology
Co., Ltd.
Shenzhen, China

Bin Wang
School of Environment and Resource
Southwest University of Science
and Technology
Mianyang, China

Yungui Li
School of Environment and Resource
Southwest University of Science
and Technology
Mianyang, China

Fengyu Wang
Yunnan Hexu Environmental Technology
Co., Ltd.
Shenzhen, China

ISBN 978-981-99-5908-2   ISBN 978-981-99-5906-8   (eBook)
https://doi.org/10.1007/978-981-99-5906-8

Jointly published with China Environment Publishing Group Co., Ltd.
The print edition is not for sale in China (Mainland). Customers from China (Mainland) please order the
print book from: China Environment Publishing Group Co., Ltd.

This Springer imprint is published by the registered company Springer Nature Singapore Pte Ltd.
The registered company address is: 152 Beach Road, #21-01/04 Gateway East, Singapore 189721,
Singapore

Paper in this product is recyclable.

Foreword

Water resources protection is the top priority of environmental protection, which needs to be achieved through sustainable, comprehensive, and efficient water pollution control. In recent decades, domestic sewage treatment in large and medium-sized cities has developed comprehensively and steadily and has achieved gratifying results. The key task has gradually shifted to towns and villages. In recent years, although rural sewage treatment has entered a rapid development stage, there are still many difficulties and problems. In order to retain the green mountains and rivers, rural sewage treatment must be strengthened to meet the difficulties.

As early as half a century ago, with the process of urbanization, developed countries have gained rich experience in decentralized domestic sewage treatment, but China's rural land area is vast, the population is large, the regional characteristics and living habits are different, coupled with the small scale of domestic sewage treatment, large changes in water quality and quantity, uneven level of operation and maintenance technology, and limited funds. It is necessary to explore the way of rural sewage treatment in line with China's national conditions.

Fortunately, the top-level design at the national level, such as the Action Plan for Water Pollution Prevention and Control, the Strategic Plan for Rural Revitalization, and the Law of the People's Republic of China on Promoting Rural Revitalization, tends to be perfect, and the pollutant discharge standards for rural domestic sewage treatment plants in 31 provinces, autonomous regions, municipalities directly under the Central Government, and special administrative regions have been promulgated one after another. Equipment standards, evaluation and certification rules, operation and maintenance technical regulations, and local technical guidelines for rural domestic sewage treatment have also been formulated, laying a good foundation for rural sewage treatment.

After nearly 20 years of exploration and practice, China's rural sewage treatment has gradually formed a treatment technology with activated sludge process and biofilm process as the core and produced a number of integrated treatment devices with simple installation, operation and maintenance, and high standardization.

The author of this book has been specializing in rural domestic sewage treatment for ten years and is one of the leading enterprises in rural sewage treatment. On the basis of summing up experience and lessons, he has written this book. This paper not only introduces the characteristics, treatment process, and management technology of rural domestic sewage, but also expounds the practical and operable integrated treatment technology route from the aspects of technical stability, equipment economy, and sustainability of operation and maintenance. In addition, the design and manufacture of integrated equipment, automatic control and cloud management platform, as well as typical application cases were shared.

This book is of great significance to the promotion of typical technologies and the construction of rural sewage treatment plants.

As a fellow traveler in the field of environmental protection and a witness of the continuous growth of the environment, I am very pleased to see the publication of this book. It is believed that the publication of this book can provide experience and reference for rural sewage treatment and contribute to the green "one belt and one road" project.

Beijing, China   Shijun Hang

Preface

The Belt and Road (B&R) Initiative is a remarkable move in implementing the concept of building a community with a shared future for mankind, contributing Chinese wisdom and schemes to the world. Green development and construction have become dominant with the intensive advancement of B&R. The B&R involves many countries in the continents of Asia, Europe, and Africa, which are confronted with the needs for accelerating economic development and strengthening environmental governance for their fragile ecosystems. A win-win situation can be achieved in environmental protection and economic development by green B&R, contributing to the construction of global ecological civilization on the premise of establishing mutual trust between China and B&R countries.

Technological and business communities in China have conducted extensive exchange and cooperation with B&R countries in order to address their prominent water security and environmental problems. China's water treatment industry and products have emerged on the world stage with their technological advantages. Chinese enterprises, for instance, have carried on many environmental protection supporting projects in B&R countries, and water treatment plants made in China have been exported to Southeast Asia, Central Asia, the Middle East, Africa, and other regions in large quantities.

Combining sewage pretreatment, biological treatment, sedimentation, and disinfection, the integrated sewage treatment equipment, featuring compact structure, small footprint, short construction period, high treatment efficiency, and economical rationality, is particularly suitable for decentralized domestic sewage treatment in rural areas without pipe networks. First applied in Japan, the integrated sewage treatment equipment has recently achieved rapid growth in China as China has vast territory and witnesses continued urbanization. The relevant technological R&D, practice, and promotion experience accumulated by Chinese enterprises can offer significant reference to other developing countries and regions in terms of sewage treatment in rural areas.

Since its founding in 2012, HEXU Environment remains committed to the R&D and production of integrated sewage treatment equipment and now has four decentralized sewage treatment product lines named Beisi, Naisi, CHTank, and FREETANK for towns, villages, and connected/single households as well as intelligent management software, IoT control hardware, and cloud platform information system. HEXU Environment has maintained its global leading position in the field of sewage treatment in rural areas through iteratively updating products and technologies from in-depth technical exchanges with overseas outstanding enterprises and Chinese scientific research institutes. HEXU has also established an overseas market department with products launched into the international market.

As HEXU marks its tenth anniversary, this book, published together with our long-term partner the School of Environment and Resources of Southwest University of Science and Technology and the Sichuan International Science and Technology Cooperation Base of "Low-Cost Sewage Treatment Technology", aims to share our achievements and experience, gives the role of the integrated treatment technology and equipment in the sewage treatment in rural areas under B&R into better play, and provides theoretical and technical references for the continued improvement of the integrated sewage treatment process system.

The book consists of five chapters. To be specific, Chap. 1 introduces characteristics, environmental risks, and management styles of domestic sewage in rural areas. Chapter 2 illustrates the collection and treatment approaches for rural domestic sewage. On this basis, the integrated sewage treatment process of high efficiency is presented in Chap. 3, including the A^3/O-MBBR process, the modified Bardenpho-MBBR process, the multi-stage A/O-MBBR process, the multi-stage A/O biological contact oxidation process, SND-type biological contact oxidation process, A/O-MBR process, and membrane aeration biofilm reactor. Further, Chap. 4 studies the design concept, manufacturing requirements, automatic control, and the cloud management system for the integrated sewage treatment equipment. Chapter 5 discusses the typical application cases of the integrated treatment process in multiple regions.

Authors and editors of this book include Wensheng Li, Yungui Li, Jianmin Zhang, Fengyu Wang, Bin Wang, Peng Deng, Ranrong Liu, Xiaohui Lv, Yuqiong Wang, Jing Liao, Qingsong Wu, Jun Yang, Hua Du, Songtao Zhang, Bo Zeng, Honglei Sun, Dingchuan Xie, Weihua Chen, Jingguo Ding, and Zheng Lv. Our deep appreciation also goes to the strong support given by China Environmental Science Press and the assistance from the China Academy of Engineering Physics, Nisshinbo Group, and other partners as well as the help of Haibo He, Xin Zhang, Gangquan Zhou, Tingting Yang, and Liutang Yang in writing, proofreading, and publishing this book.

We would also love to express sincere gratitude to related authors of the research findings and application experience that are referred to in this book. Due to restrictions in various conditions, there would be inevitable omissions and inaccuracies in the book, and readers are welcome to leave comments.

Shenzhen, China   Wensheng Li
Mianyang, China   Yungui Li
Shenzhen, China   Jianmin Zhang
Shenzhen, China   Fengyu Wang
Mianyang, China   Bin Wang
May 2022

Contents

Chapter 1
Characteristic and Management of Rural Domestic Sewage

1.1 Sources and Characteristics of Domestic Sewage in Rural Areas

1.1.1 Source and Composition

The domestic sewage (from bath, laundry, kitchen, toilet) flushing is the main source of domestic sewage in rural areas, with their proportions associated with climatic conditions, living standards, and living habits. According to the water quality of sewage, domestic sewage (from bath, laundry, kitchen, toilet) is classified as gray water, while the toilet-flushing sewage generated from excretion and feces flushing is known as black water (Paulo et al., 2013).

Compared with gray water, black water cannot be treated easily due to its high concentration of pollutants, but it features high-value resource utilization as a potential resource. Domestic sewage in rural areas is composed of complicated pollutants, including solid matters, organic matters, nitrogen compounds, phosphorus compounds, animal and vegetable oils, anionic surfactants, drugs, personal care products, bacteria, viruses and insect eggs. SS, COD, BOD, TN, NH_3–N, and TP are critical water quality indicators that should be highly concerned in rural sewage treatment.

Specifically, SS is the physical measure of suspended solid inorganic particles (silt and clay), organic particles, and their loaded microorganisms for detecting water pollution. Note that the high concentration of SS might lead to murky water and reduced water clarity, consequently affecting the respiration and metabolism of aquatic organisms and destroying the aquatic ecosystem. More than that, organic suspended particulate matters are prone to anaerobic fermentation upon deposition, resulting in deterioration of water quality (Palleiro et al., 2013).

COD and BOD are the main evaluating indicators for the level of organic pollutants in water bodies. Excessively high COD and BOD contents indicate high contents of

© The Author(s) 2024
W. Li et al., *Integrated Treatment Technology of Rural Domestic Sewage*,
https://doi.org/10.1007/978-981-99-5906-8_1

reductive organics and large DO consumption, prone to water environment problems such as black and odor water. Compared with the inflow of municipal sewage treatment plants, rural domestic sewage has higher organic contents (COD reaching up to 500 mg/L) with favorable biodegradability (BOD_5/COD, 0.45–0.55), and the organics can be effectively removed by biochemical treatment (Fan et al., 2021; Xu et al., 2012).

TN is the sum of different nitrogen forms, including inorganic nitrogen (NH_3–N, NO_2^-–N, NO_3^-–N, etc.) and organic nitrogen (proteins, amino acids, organic amines, etc.). Organic nitrogen and NH_3–N are primary nitrogen forms in rural domestic sewage, and both are unstable. The former can be converted into NH_3–N through ammonification, and the latter can be converted into NO_2–N via nitrification and further oxidized to NO_3–N. The consumption of water dissolved oxygen in the conversion process is a cause of deteriorated water quality (Fu et al., 2006; Ma et al., 2019). Moreover, excessively-high concentrations of NH_3–N, NO_2–N, and NO_3–N have direct or indirect toxic effects on aquatic ecosystems, resulting in hazards such as eutrophication. Hence, TN and NH_3–N are technical indicators that are essential for managing the discharge of domestic sewage in rural areas. Rural domestic sewage treatment has high nitrogen load and faces higher technical difficulties in treatment process design and engineering operation and maintenance than that in urban areas.

TP, including dissolved total phosphorus and solid total phosphorus, is the sum of various phosphorus species in water. Human feces, food residues, and phosphorus-based detergents are primary sources of TP in domestic sewage. Like nitrogen, phosphorus is an essential element for biological growth and an indicator of nutrient pollution in water (Powers et al., 2016; Withers et al., 2009). Excess nitrogen and phosphorus in water may lead to the excessive multiplication of algae, giving rise to eutrophication. Phosphorus treatment is a ticklish problem for rural domestic sewage treatment since the biological phosphorus removal efficiency is not stable enough and the chemical phosphorus removal process is defective in difficult operation and maintenance and high price.

1.1.2   Main Characteristics

Rural domestic sewage features a large total amount of discharge, extensive discharge area, great fluctuation in water quality and quantity, remarkable regional characteristics of discharge, low treatment rate, and great potential for recovery.

a.   Large Amount of Discharge in Total

A majority of B&R countries are developing countries, which have a large rural population and a large total amount of domestic sewage discharge. There are 634,000 administrative villages and 22,000 towns in China, generating an annual discharge of domestic sewage exceeding 20 billion tons, accounting for more than half of the national total amount. 4.9962 million tons of COD, 245,000 tons of NH_3–N, 446,500 tons of TN, and 36,900 tons of T are discharged from rural domestic sewage,

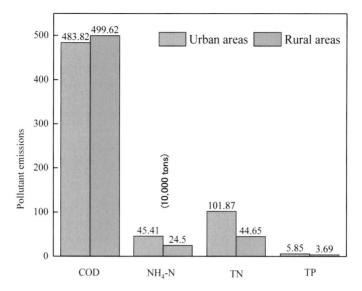

Fig. 1.1 Comparison of discharges of main pollutants in urban and rural sewage in China (second national survey of pollution source)

accounting for 50.8, 35.0, 30.5, and 38.7% of the total discharge of domestic sewage pollutants (including the discharge of domestic source water pollutants in urban areas), according to China's Second National Survey of Pollution Source shown in Fig. 1.1.

b. **Extensive Discharge Area**

The scattered living area of rural residents and a wide range of domestic sewage discharge areas are important sources of water source pollution due to defective sewage collection pipe networks and difficulty in concentrated treatment. Domestic sewage in rural areas is commonly treated with the following processes.

(i) leveraging treatment. Those villages near the urban sewage centralized collection system get access to the municipal and enterprise sewage pipe network for treatment; (ii) independent treatment. Rural areas with developed economies and management build their own domestic sewage treatment system (such as the integrated treatment equipment, and constructed wetlands) to discharge sewage upon treatment; (iii) simple treatment or direct discharge. In cases of centralized living, open ditches or underdrains, or simple facilities such as roadside ditches are adopted to collect and discharge sewage; in cases of scattered living, direct discharge is adopted (Jin, 2021; Li & Yu, 2021).

c. **Great Fluctuation in Water Quantity and Quality**

The quantity and quality of rural domestic sewage fluctuate greatly under the impacts of factors such as daily routine, and periodic population flow. With the great mobility of the rural population, the discharge of domestic sewage is significantly increasing during holidays when a large number of people return home. Evident seasonal variation is presented in the discharge of rural domestic sewage. As residents have large demands for bathing in summer, leading to increased discharge of washing-up sewage, sewage discharge then is primarily characterized by large water volume and low concentration (COD, TN, TP). On the contrary, it is characterized by small water volume and high discharge concentration in winter. Similar discharge characteristics can be observed in spring and autumn, showing intermediate water quality and quantity. In addition, dilution by rainfall is also an important influencing factor. Lu et al. (2020) found that the average concentration of different pollutants in the rainy season is only 20–50% of that in the dry season upon studying villages in the Erhai Basin of Dali.

d. Remarkable Regional Characteristics of Discharge

Sewage discharge in rural areas is closely bound up with climatic conditions and economic development levels, presenting remarkable regional characteristics. Different from residents in southeast China, rural residents in northwest China with less domestic water consumption under the impacts of the dry climate, particularly low frequency of washing up, discharge a smaller amount of sewage.

Liang et al. (2011) found that for a rural family of five, the water consumption for showering in South and North China is 70–140 L/(person d) and only 25 L/(person d) in summer, respectively. Also, for the water consumption in the kitchen, the water consumption of the farmer families in North China and Zhejiang is 50–70 L/d and 80–130 L/d, respectively.

Domestic sewage discharge in rural areas is positively correlated with the economic level of regional villages. With the high-speed economic growth, the output of rural domestic sewage is also increased together with the accelerated modernization of farmhouses and villages, the access of tap water to the farmhouse, and improved quality of rural life (such as the convenient use of modern kitchen appliances and bathroom products). Besides, the industrial structure also affects the discharge of domestic sewage in rural areas. The discharge of sewage from villages that are tourist destinations fluctuates remarkably in peak and low peak tourist seasons (Liang et al., 2011; Lu et al., 2020).

e. Low Treatment Rate

The rate of domestic sewage treatment in rural areas of the B&R developing countries is low due to economic and technical constraints. The coastal developed areas in East and Southeast China with favorable economic growth took the initiative in rural sewage treatment and heavily invested in sewage treatment, achieving a complete rural environmental treatment infrastructure and full coverage of rural sewage treatment. Conversely, a low treatment rate of rural sewage can be observed in the central and western parts of China due to insufficient investment in rural environmental governance infrastructure as well as influence of living habits and natural conditions.

f. **Great Potential for Recovery**

Rural domestic sewage is large in volume and extensive in distribution, which can be changed into large numbers of secondary water resources upon effective treatment for efficient onsite utilization. Besides, nutrient substances, such as N/P, abundant in rural sewage have a high value in recycling. Recycling N/P from rural domestic sewage (sludge) as fertilizer can realize the efficient utilization of nutrients, and avoid eutrophication by stopping nitrogen-containing compounds from entering the water. Moreover, carbon-rich domestic sewage has the potential for energy development. The sludge generated from the domestic sewage treatment has high chemical energy storage upon stabilized treatment, with the heat value reaching up to 32 MJ/kg (Hu et al., 2016; Li et al., 2020). It can be seen that the resource utilization of rural domestic sewage has great potential, which is especially significant for arid and semi-arid regions with scarce resources.

1.1.3 Environmental Risks

A range of environmental risks can be incurred after discharging rural domestic sewage without effective treatment/recycling, including (a) Contamination of drinking water sources. Discharging untreated domestic sewage or sewage below standard into surface water may contaminate drinking water sources, resulting in disease transmission. (b) Contamination of soil and groundwater. Rural domestic sewage discharged without effective treatment in areas with low groundwater levels may spark a high risk of groundwater contamination with indicators such as *Escherichia coli* exceeding the standard. (c) Disrupted balance of aquatic ecosystems and reduced stability and diversity of aquatic organisms affect fish survival and fishery production. (d) Black and odorous water. Continuous fermentation of organic matters in rural domestic sewage may generate odorous substances such as hydrogen sulfide, mercaptan, and ammonia, as well as black substances such as iron sulfide and manganese sulfide, resulting in black-odor water and loss of its function and affecting the landscape and human health. (e) Water eutrophication. The excessive multiplication of aquatic organisms would lead to a rapid drop in water transparency and dissolved oxygen, a sharp increase in pollutant indicators, and deteriorated water quality. (f) Breeding of mosquitoes and flies. Accumulation of domestic sewage provides an incubator for the reproduction of pests, such as mosquitoes and flies.

1.2 Management of Domestic Sewage in Rural Areas

Management policy is the driving force for addressing ecological and environmental problems. But the management of rural sewage is less developed than that of urban sewage. The United States and Japan are among the earliest countries to study rural sewage management. China, as a developing country and the second largest economy in the world, has also accumulated rich experience in rural sewage management.

1.2.1 United States

The U.S. began to address the problem of rural domestic sewage treatment as early as the middle of the 19th century and built domestic sewage treatment plants. Its rural decentralized sewage treatment technology evolved as outdoor toilets → sewage sumps → septic tanks → decentralized sewage treatment system. Complete systems concerning the regulatory system and financial subsidy system as well as technology and operation models have been established (Fan et al., 2009; Wen, 2016).

The same legal system was applied in rural and urban sewage treatment in the United States in the 1960s, highlighting the rural households' independent management of domestic sewage; and rural domestic sewage treatment plants achieved full coverage in the 1980s. The treatment of non-point source pollution was included in the *Water Quality Program* in 1987 to request that programs and project funding should be established for decentralized sewage treatment in all states. But the investigation in 1997 by EPA found that the national decentralized sewage treatment system was not operated satisfactorily. Accordingly, a series of guidance documents on decentralized sewage treatment and management were released by EPA to strengthen the treatment of rural domestic sewage and determine jurisdiction and rights over the decentralized sewage systems in accordance with the management capacity and jurisdiction of local governments. Moreover, five operating modes (owner independent mode, agreement maintenance mode, licensed operation mode, centralized running mode, and centralized operation mode) with strengthened centralized management were proposed for decentralized sewage treatment in the *"Guidelines for the Management of Decentralized Treatment System"* issued in 2003. Besides, EPA has also established and governed various programs associated with the management of decentralized sewage treatment systems, such as the Water Quality Standard Project, the Total Maximum Daily Load Program, the Non-Point Source Management Program, the National Emission Reduction System Program, and the Water Conservation Program. Meanwhile, effective financial guarantee systems such as the state rolling fund programs provided by states for significant sewage treatment and environmental protection programs were also established for domestic sewage treatment in rural areas.

The complete decentralized sewage policy system, the multi-dimensional operation system, and the strong financial support system are essential for the successful treatment of rural domestic sewage in the U.S. This experience would be helpful for the rural domestic sewage treatment in B&R countries, which include:

i.   Domestic sewage treatment policies for rural areas should be flexibly formulated according to local conditions as well as the environmental needs, geographical conditions, economic development level, sewage discharge status, and sewage collection system.

ii. A sound market mechanism and management mechanism of rural domestic sewage treatment should be established to give the roles of various forces such as the government, non-governmental organizations, and enterprises into full play.

iii. An effective capital guarantee system should be built to expand the financing channels for domestic sewage treatment in rural areas.

1.2.2 Japan

With a small territory, large population and shortage of water resources, Japan has a significant demand for rural domestic sewage treatment. Its primary treatment processes are Johkasou treatment and centralized treatment, and has established a sound legal system, financial subsidy system, and operation mode for domestic sewage treatment. Overall, the comprehensive treatment level of rural domestic sewage in Japan is second to none among the developed countries.

Japanese enterprises began to apply Johkasou technology and facilities in rural manure treatment in the 1960s. The *"Construction Benchmark Law"* was promulgated by the Japanese government in 1969 for regulating and developing the market. In 1977, the rural domestic sewage treatment plan was launched, and the integrated water treatment equipment (Johkasou) with the biofilm process or activated sludge process was developed with reference to the urban sewage treatment technology. The *Johkasou Law* was officially enacted in 1983 to govern the decentralized treatment of rural sewage, and has become the principal legal basis for the treatment of rural domestic sewage in Japan. Further, a subsidy system was established for the "consolidated Johkasou setting and reconditioning business" in 1987, and the subsidy system for the "domestic drainage treatment business in specific areas" has been implemented since 1994 (Zhao et al., 2018).

The Japanese government, with the *Johkasou Law* as the core, has established a complete law and policy system for rural domestic sewage treatment, proposed requirements for the treatment and discharge of domestic sewage at different scales, established technical guidance and market supervision for the construction, operation, and maintenance of Johkasou facilities, strengthened the fund guarantee system for facility construction and operation and propelled the standard, professional and large-scale development of the Johkasou industry. On this basis, a sound policy environment has been built for the treatment of rural domestic sewage in Japan. Furthermore, the *"Rules of Enforcing the Johkasou Law"*, *"Structural Standards and Explanations of Johkasous"*, and *"Design Guidelines for Drainage Facilities in Agricultural Villages"* have been also introduced to support the implementation of the *Johkasou Law*. Meanwhile, local regulations and standards have been also unveiled by various prefectures to identify the responsibilities and obligations of

relevant national departments, local governments, facility users as well as operation and maintenance agencies and personnel, laying a foundation for the orderly
advancement and scientific management of rural domestic sewage treatment (Chen
et al., 2019).

Rural domestic sewage treatment is fully covered in Japan, realizing the efficient
utilization of resources. Specifically, the treated water in roughly 80% of the area is
recycled as agricultural water, and approximately 71% of the sludge generated from
the rural drainage facilities is recycled through farmland restoration. The collected
biomass, for example, is used as fertilizers after being mixed and fermented with
accessory materials (such as rice husks).

1.2.3 China

Compared with the United States and Japan, rural sewage treatment in China starts
quite late in the early 21st century, and has experienced its infancy, development
state, and rapid development stage.

Stage 1, infancy (2005–2008). China began to highly focus on rural environmental
protection and attempted to guide industrial development via the formulation of policies. The State Council held the first National Rural Environmental Protection Work
Conference in 2008, kicking off the comprehensive treatment of decentralized sewage
in rural areas. Meanwhile, the Ministry of Finance and the Ministry of Ecology and
Environment established a special fund for rural environmental protection to support
rural domestic sewage treatment and refuse disposal.

Stage 2, development stage (2009–2015). Focusing on policy discussion,
funding support, and demonstration base construction, the government issued the
"*Implementation Plan on the Accelerated Resolution of Outstanding Rural Environmental Problems with Awards to Promote Governance*", the "*Rural Domestic Pollution Control Technology Policy*", the "*Environmental Protection Act of the People's
Republic of China*", and the "*Action Plan for Water Pollution Control*". The environment of 60,000 administrative villages has been comprehensively improved from
decentralized pilots to centralized contiguous coverage through the implementation of promoting governance with awards. In this way, a number of demonstration
cases were accumulated in environmentally-sensitive and economically-developed
areas. Also, decentralized sewage treatment enterprises have been developed rapidly
(HEXU Environment was founded in 2012), and they have developed a range of rural
sewage treatment technologies and processes as per national conditions in China.

Stage 3, rapid development stage (2016–). It is a stage characterized by improved
policies and mechanisms as well as the vigorous boost of regional integrated services.
The Ministry of Housing and Urban-Rural Development issued the "*Technical
Standard for Domestic sewage Treatment Engineering in Rural Areas*" in 2019 to
provide guidelines for the technological development of the industry. Water pollutant discharge standards for local rural domestic sewage treatment plants have been
established in 31 provinces, municipalities directly under the Central Government,

and autonomous regions across the country as per actual conditions by 2020, meeting the requirements of decentralized sewage treatment in rural areas. The Ministry of Finance issued the *"Management Measures for Rural Environmental Improvement Funds"* in 2021 to provide fund support for rural sewage treatment nationwide, and the China Council for the Promotion of International Trade released the *"Standards for Small Treatment Equipment for Domestic sewage"*, *"Evaluation Specifications for Small Treatment Equipment of Domestic sewage"*, and *"Technical Regulations for the Operation and Maintenance of Domestic sewage Treatment plants in Villages"* successively. The environment of 130,000 administrative villages has been comprehensively improved under the implementation of the *"Action Plan for Agricultural and Rural Pollution Control"* during the 13th Five-Year Plan period, and the environment of 80,000 administrative villages will be comprehensively improved during the 14th Five-Year Plan period.

China has established complete legal systems, financial subsidy systems, and operation models, achieving significant improvement in the comprehensive treatment level of domestic sewage in rural areas. China's experience accumulated in rural sewage treatment over the last two decades by drawing the advanced treatment concepts of the United States and Japan can be used as a reference for our B&R partners.

References

Chen, Y., Yu, Q., & Jia, X. (2019). Improving policies and mechanisms of rural sewage treatment in China with reference to Japanese purification tank law. *Chinese Journal of Environmental Management, 11*(2), 14–17.

Fan, B., Wu, J., Liu, C., et al. (2009). Organization, administration and illumination of rural sewage treatment in USA and Japan. *China Water and Wastewater, 25*(10), 6–10, 14.

Fan, Z., Liang, Z., Luo, A., et al. (2021). Effect on simultaneous removal of ammonia, nitrate, and phosphorus via advanced stacked assembly biological filter for rural domestic sewage treatment. *Biodegradation, 32*(4), 403–418.

Fu, R., Yang, H., Gu, G., et al. (2006). Nitrogen removal from rural sewage by subsurface horizontal-flow in artificial wetlands. *Technology of Water Treatment, 1*, 18–22.

Jin, M. (2021). Analysis on the present situation and treatment of rural domestic sewage in Yuzhong, Gansu. *Gansu Science and Technology, 37*(4), 4–7.

Hu, M., Fan, B., Wang, H., et al. (2016). Constructing the ecological sanitation: A review on technology and methods. *Journal of Cleaner Production, 125*, 1–21.

Li, B., & Yu, J. (2021). Analysis of rural domestic sewage treatment in Zhaohe River basin of Hefei City. *Resources Economization and Environmental Protection, 8*, 76–78.

Li, Q., Wu, J., Wu, Z., et al. (2020). Domestic sewage resource recovery technology and relevant visual analysis of domestic sewage resource. *Environmental Science and Technology, 43*(6), 223–229.

Liang, H., Wu, J., Wei, Y., et al. (2011). Investigation and analysis of rural wastewater discharge characteristics in three typical areas of China. *Chinese Journal of Environmental Engineering, 5*(9), 2054–2059.

Lu, H., Zhang, X., Wu, C., et al. (2020). Investigation and analysis of rural domestic water use and sewage discharge in Dali Erhai Lake basin. *Environmental Impact Assessment, 42*(6), 6.

Ma, L., Liu, W., Tan, Q., et al. (2019). Quantitative response of nitrogen dynamic processes to functional gene abundances in a pond-ditch circulation system for rural wastewater treatment. *Ecological Engineering, 134,* 101–111.

Palleiro, L., Rodriguez-Blanco, M. L., Taboada-Castro, M. M., et al. (2013). The influence of discharge, pH, dissolved organic carbon, and suspended solids on the variability of concentration and partitioning of metals in a rural catchment. *Water Air and Soil Pollution, 224*(8), 1651.

Paulo, P. L., Azevedo, C., Begosso, L., et al. (2013). Natural systems treating greywater and blackwater on-site: Integrating treatment, reuse and landscaping. *Ecological Engineering, 50,* 95–101.

Powers, S. M., Bruulsema, T. W., Burt, T. P., et al. (2016). Long-term accumulation and transport of anthropogenic phosphorus in three river basins. *Nature Geoscience, 9*(5), 353.

Wen, Y. (2016). *A study on wastewater treatment system in Chinese typical villages.* Tsinghua University.

Withers, P. J. A., Jarvie, H. P., Hodgkinson, R. A., et al. (2009). Characterization of phosphorus sources in rural watersheds. *Journal of Environmental Quality, 38*(5), 1998–2011.

Xu, J., Xu, L., Jiang, J., et al. (2012). Treatment effect of combined ecological and biochemistry wetlands on disposing rural domestic sewage. *Journal of Agro-Environment Science, 31*(9), 1815–1822.

Zhao, F., Jia, X., & Li, D. (2018). The experience of rural sewage treatment in Japan as a reference for China's rural revitalization strategy. *World Environment, 2,* 19–23.

Chapter 2
Overview of Rural Domestic Sewage Treatment Technology

2.1 Collection of Domestic Sewage in Rural Areas

2.1.1 Collection Modes

Collection, as the premise of treatment, is one of the important factors affecting the efficiency of sewage treatment. According to the experience of China, Japan, Europe, the United States, and other countries, decentralized collection, centralized collection, and nanotube collection are three basic modes for collecting rural domestic sewage. Various factors such as local population distribution, sewage quantity, economic development level, environmental characteristics, climatic conditions, topography, local drainage system, and current status of drainage pipeline network should be comprehensively considered before selecting a sewage collection mode. Moreover, the collection system should be planned based on local conditions and the distribution of villages and farmer households. Long-distance drainage pipes should be avoided.

To be concrete, decentralized collection characterized by saving the investment in pipe network as well as simple and fast construction, is applicable for areas not suitable for pipe networks due to complex terrain, scattered village distribution, and low population density. Decentralized collection can be divided into decentralized collection and treatment as well as decentralized collection and centralized treatment, according to different treatment modes (Table 2.1 and Fig. 2.1). More precisely, indoor sewage collection pipes and single household/connected household integrated sewage treatment equipment are prerequisites for decentralized collection and treatment. Indoor sewage collection pipes, septic tanks, and collection tankers for periodic clearing are required for decentralized collection and centralized treatment with less investment due to the simple structure and low cost of the septic tanks. But it is worth noting that the accumulation rate and clearing frequency of toilet sewage in septic tanks are high in areas with a high penetration rate of flush toilets, which might be beyond the capacity of the collection and transportation system (Wang, 2021a).

© The Author(s) 2024
W. Li et al., *Integrated Treatment Technology of Rural Domestic Sewage*,
https://doi.org/10.1007/978-981-99-5906-8_2

Table 2.1 Comparison of collection of domestic sewage in rural areas

Collection mode		Scope of application	Supporting facilities and equipment
Decentralized collection	Decentralized processing		Indoor sewage collection pipe, single household/ connected household integrated sewage treatment equipment
	Centralized processing	The area with low population density, and complex terrain, and not suitable for laying pipeline networks	
Centralized collection		The area with high population density, high sewage volume, gentle terrain, and far from towns	Collection pipe network, lifting pump station
Nanotube collection		Areas close to cities, with favorable economic conditions and conditions for pipe network laying	Collection pipe network, lifting pump station

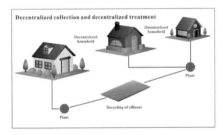

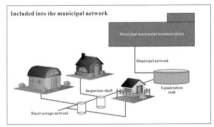

Fig. 2.1 Collection of domestic sewage in rural areas (Li et al., 2021)

Centralized collection refers to the collection of sewage by the pipe network to the nearby sewage treatment plants from farmer households within a certain range or a village, applicable for areas away from towns with gentle terrain, compact settlement, or large resident population. Where possible, a pipe network diversion system should be performed. In such a model, sewage collection, treatment, and discharge are conducted according to the principle of proximity. In other words, treated sewage can be discharged into the nearest river, or directly leveraged as farmland irrigation. The difficulty in engineering construction of centralized collection is low, but collection pipe networks and lifting pump stations are needed (Wang, 2021a, 2021b).

Nanotube collection refers to the collection of domestic sewage from towns and surrounding villages by branch sewers and directly incorporated into urban main sewers, and finally treated in municipal sewage treatment plants. It features convenient management, low investment, quick return, high collection efficiency, and significant treatment effect. Its application is limited by geographical restrictions with collection pipe networks and lifting pump stations required (Li et al., 2021).

2.1.2 Collection Process

Gravity collection is the main approach leveraged for collecting rural domestic sewage. Gravity collection is a major process used in collecting rural domestic sewage because of its superiority in low investment, low energy consumption, and simple operation and maintenance (Table 2.2 and Fig. 2.2). But gravity collection may result in problems such as difficult construction, pipelines susceptible to plugging, and sewage leakage in complex terrain. In that case, the vacuum negative pressure processing method can be adopted to collect domestic sewage in areas with difficulty in laying gravity pipe networks. It is a process that creates a vacuum in the drainage pipeline via the vacuum equipment and provides the power to convey liquid using the air pressure difference, as shown in Table 2.2 and Fig. 2.2 (Ye et al., 2021). Featuring small pipe diameter, shallow buried depth, incline capacity, and non-clogging, the above process is applicable for collecting domestic sewage under complicated geographical conditions. Although it has been applied at home and abroad, it is rarely applied in rural areas as a whole for the large investment in the vacuum collection system and the difficulty in operation and maintenance (Islam, 2017).

2.2 Domestic Sewage Treatment in Rural Areas

Compared with cities, villages and towns feature dispersed population distribution, complex terrain, and a lack of complete pipe networks, making it more difficult for sewage collection and treatment (Deng & Wheatley, 2016).

The quality and quantity of domestic sewage differ due to great differences in economic and social conditions, population factors, and customs under a diversity of geographical, climatic, and ecological conditions in rural areas worldwide. Moreover, different treatment processes selected as the sewage treatment terminal may lead to a big difference in infrastructure cost, treatment effect, and the complexity of operation and maintenance management. Evidently, a reliable and safe sewage treatment process with favorable treatment effect, low energy consumption, simple

Table 2.2 Comparison of gravity collection and vacuum collection

	Scope of application	Pipe network laying	Supporting equipment	Engineering investment	Operation management
Gravity collection	The area with flat terrain and large-scale pipe network	Deeply buried below the freezing line; limited by terrain and burial depth; a lifting pump station is required for long-distance transportation	Few	Low	Simple
Vacuum collection	The area with difficulties in laying gravity pipe network, large fall in terrain, and small pipe network scale	Shallowly buried above the freezing line; the pipeline network should be sealed, and the pipeline laying is flexible without limitations of terrain and burial depth; no lifting pump station is needed	Many	High	Complex

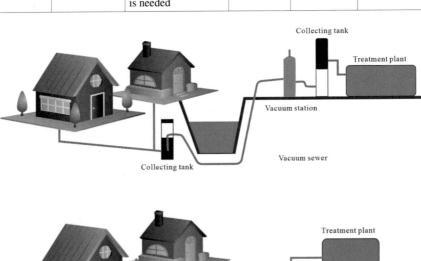

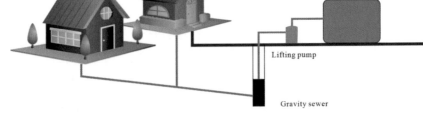

Fig. 2.2 Gravity collection and vacuum collection (Li et al., 2021)

operation and maintenance management, and up-to-standard discharge should be adopted with full consideration of local conditions rather than using a single sewage treatment process.

In this section, common processes in rural sewage treatment are discussed from the aspects of physical and chemical processes, biofilm process, activated sludge process, and natural biological treatment process. Moreover, septic-tank and MBR are commonly used in rural sewage treatment. The septic tank and MBR are studied separately for them, strictly, are neither part of the biofilm process nor the activated sludge process. Then, the rural sewage treatment selected for B&R is discussed based on the experience of rural sewage treatment in China.

2.2.1 Physical and Chemical Processes

Grid, grit chambers, equalization tanks, chemical phosphorus removal, and disinfection are common physical and chemical processes in rural domestic sewage treatment.

a. **Grid**

Grid is a simple filter device typically used as the first session for sewage treatment in the pretreatment process, which is set in front of sewage treatment plants or pumping stations.

Grid is leveraged to efficiently block and collect coarse suspended solids or floating solids such as leaves, entanglement, and solid waste in sewage, which can prevent the blocking of subsequent pipeline valves or water pumps and lay a foundation for subsequent biochemical treatment processes (Zhang, 2015).

There are artificial grids and mechanical grids, which are selected as per the required slag amount to be removed. The grid can be split into the coarse grid, fine grid, and refined grid as per the gap of the grid bar. The coarse grid with a gap of 26–40 mm is typically selected when grid slag is removed manually; and when grid slag is removed mechanically, the gaps of 15–25 mm, 3–10 mm, and 1–2 mm will be selected for the coarse grids, fine grids and refined grids, respectively. A fine grid, in general, is set before MBR (Membrane Bio-Reactor) process to prevent hair from wrapping film components.

b. **Equalization Tank**

Great changes in the water quality and quantity are one of the characteristics of rural sewage, which might seriously affect the biochemical processing at the back end with a significant change in the pollutant load of sewage. To this end, a equalization tank is normally set up at the front end of the sewage treatment plant to homogenize the quantity and quality of raw water. In this way, it ensures that biochemical treatment can be conducted at the back end under stable inflow conditions, degrade some of the organic matter, and improve the shock resistance and processing effect of the whole system (Liu, 2019). Besides, a liquid level controller is generally set in the equalization tank to adjust the water volume by controlling the start and stop of the lifting pump. At the same time, a stirring apparatus can be also installed in the equalization tank to adjust water quality.

The effective volume of the equalization tank should be determined based on factors such as the treatment scale, and the fluctuation of water quantity and quality, and the hydraulic retention time shall not be less than 12 h. Liner should be performed on the wall and bottom of the tank together with odor-resistant and explosion-proof measures. The sewage treatment plant should conform to the peak flow requirements without overflow in cases where equalization tank is not applicable.

c. Chamber

The grit chamber can be set up as required to remove solid particles with a high density such as mud and sand in a simple and effective manner (Zhang, 2015). In general, it is set before the pumping station and secondary secondary sedimentation tank to reduce the wear of the water pump, prevent the blockage of the pipeline, increase the content of organic components of the excess sludge produced by the sewage treatment plant, and improve the value of sludge as a fertilizer for resource utilization.

d. Secondary Sedimentation Tank

The main function of the secondary sedimentation tank is to achieve solid–liquid separation by removing suspended solids in sewage with the help of gravity precip-itation. It was used independently in sewage treatment at the early stage, presenting limited treatment effect, and now is used in combination with biological treatment processes. The secondary sedimentation tank is composed of primary and secondary sedimentation tank, with the former rarely applied in rural sewage treatment.

The primary secondary sedimentation tank is set before the biological treatment unit and mainly utilized to remove the suspended solids (SS) dominated by organic matters in the raw water, and, at the same time, improve the operating conditions of the biological treatment unit in the later stage and reduce the pollutant load. The secondary sedimentation tank is set behind the biological treatment unit to precipitate and separate activated sludge or biofilm exfoliation, and achieve certain sludge concentration. It is an integral part of the biological treatment system.

The SS removal efficiency of the secondary sedimentation tank is proportional to its size and the SS sedimentation rate, and is inversely proportional to the daily treated water volume. Hence, the surface hydraulic load (daily treated water volume/ surface area) is critical in the design of the secondary sedimentation tank.

e. Phosphorous Removal

The water quality discharged from the rural domestic sewage treatment plant is generally determined by environmental sensitivity and environmental requirements of receiving water. When the effluent quality standard limits TP discharge, the biolog-ical phosphorus removing technology will be adopted. And the chemical phosphorus removal technology will be supplemented to further treat TP when the discharge requirements cannot be satisfied by the biological phosphorus removal. Chemical phosphorus removal technology can be divided into chemical process and physical–chemical process, as shown in Table 2.3. The former includes the precipitation process and crystallization process, and the latter includes the adsorption process (Wu et al., 2019).

The precipitation process is to remove phosphorus in sewage through precipitation generated from the reaction of metal ions and phosphate radicals by adding metal salts (such as aluminum salts, iron salts, and calcium salts). This is a common process of phosphorus removal in sewage. Al^{3+} salts (such as polyaluminum chloride PAC), instead of calcium hydroxide (slaked lime), are frequently used in the treatment of rural sewage with the precipitation process. The supplement of chemical phosphorus removal in the biochemical treatment by adding chemicals can be considered in the centralized treatment of rural sewage.

As there is a risk of depleted phosphorus resources in the next few decades, phosphorus recovery and utilization are one of popular topics in the field of sewage treatment. Recovery and resource utilization of phosphorus in sewage can be achieved by removing phosphorus with the crystallization process. The representative MAP process (Magnesium Ammonium Phosphate hexahydrate), taking advantage of the reaction of magnesium salts with ammonia ions and phosphoric acid in sewage, can generate ammonium magnesium phosphate precipitation while removing nitrogen and phosphorus. Although the crystallization process is rarely used in rural sewage treatment, it will have a broad application prospect with the introduction of relevant policies on the utilization of rural sewage resources (Liu et al., 2013).

Table 2.3 Comparison of chemical phosphorus removal processes

	Phosphorus removal process	Advantages	Disadvantages
	Precipitation (such as PAC)	Adjustable operation, high stability and extensive application range	Need to configure dosing system, a large amount of sludge produced, higher treatment charge
Chemical process	Electrochemical process (such as electrolytic phosphorus removal)	Simple and adjustable operation, high stability, long operation and maintenance period	The colored water might be discharged under bad control, being affected greatly by power supply
	Crystallization (such as MAP process)	Favorable stability and phosphorus recovered	Highly affected by pH and high treatment cost
Physical–chemical process	Adsorption process (such as activated carbon, biomass, and metal oxides, etc.)	Simple device structure, simple in operation and control, and regenerated adsorbent	Less phosphorus removed

Rural domestic sewage treatment plants are distributed dispersedly with a lack of specialized operation and maintenance personnel. The electrolytic dephosphorization process has been widely used in recent years to ease the difficulty of operation and maintenance of sewage treatment plants (Sato, 2013). In electrolytic dephosphorization, the anode and cathode generate metal ions and hydrogen, respectively, with metals such as iron and aluminum as electrode plates during electrolysis; on this basis, phosphorus in sewage can be removed with the precipitation generated from the reaction of metal ions and phosphate groups and the air flotation effect of hydrogen. The operation and maintenance period can be extended to 2–3 months when double iron plate electrodes are utilized and the positive and negative electrodes are switched regularly.

Phosphorus removal by the adsorption process (such as activated carbon adsorption, oyster shell, and other biomass adsorption, iron oxide, and hydrotalcite adsorption) is to remove phosphorus through the bonding effect between the active groups of the removal agent and the phosphorus in sewage. It features fast adsorption rate, and environmentally friendly removal. But it also presents defects in low saturated adsorption capacity and complicated operation and maintenance (Wu et al., 2019). As some suggested, recently, that chemical phosphorus removal should not be used in rural domestic sewage treatment plants with a treatment scale of less than 20 m^3/d, the adsorption process has certain application prospects in rural sewage treatment.

f. **Disinfection**

Disinfection should be performed before effluent when the bacterial index is limited in the discharge standard of the sewage treatment plants. Chlorination disinfection and ultraviolet disinfection are commonly used disinfection processes in rural domestic sewage treatment under the limitation of factors such as operating costs, complexity of operation and maintenance management (Zeng et al., 2021).

Sodium hypochlorite disinfection can destroy proteins, nucleic acids, and other components in bacteria and viruses relying on the strong oxidizing properties of its hydrolyzed product hypochlorous acid (HClO). Note that a strong disinfection effect can only be achieved after reaching the chlorine breaking point when reducing substances and ammonia nitrogen in sewage are all consumed and free chlorines (HClO and hypochlorite ion) are accumulated. Sodium hypochlorite solution with a concentration of more than 10% is usually added to the clear water tanks through a metering pump for disinfection in rural sewage treatment (Zeng et al., 2021).

Chlorine dioxide (ClO_2) with a strong oxidizing property is an efficient and environmentally friendly disinfectant that can directly oxidize microbial proteins, nucleic acids, and other components in molecular form after dissolving in water (Ren, 2005). ClO_2 effervescent tablets can be selected in rural sewage treatment with effective ClO_2 content of roughly 10%, featuring convenient storage and use. A certain amount of effervescent tablets are dissolved in water, and then delivered to the clear water pond by a diaphragm metering pump for disinfection (Zeng et al., 2021).

Ultraviolet disinfection is a physical disinfection process, and the best disinfection effect can be achieved at the band near 253.7 nm. It realizes disinfection by destroying and changing the structure of DNA and RNA in microorganisms (Shiohara et al., 2005). Ultraviolet disinfection with the strengths of simplicity, convenience, high efficiency, no secondary pollution, easy management, and automation is also susceptible to the turbidity of sewage, SS, and inorganic ions (such as Fe^{3+}).

The disinfection in rural sewage treatment has received extensive attention after the worldwide outbreak of COVID-19 pandemic in 2020. Zeng et al. (2021) indicate that integrated equipment should be adopted in rural sewage treatment plants in response to the pandemic, as its independent and closed treatment unit can reduce the contact between sewage and management personnel. Detailed requirements covering the disinfection of septic tanks, tailwater, and sludge are proposed in the "*Technical Regulations for the Operation and Maintenance of Village Domestic sewage Treatment plants*" (T/CCPITCUDC-003-2021) for the operation and maintenance of sewage treatment plants during the pandemic. As pandemic prevention and control have become the new normal, rural sewage treatment plants should have the ability to conduct emergency disinfection and detection, meeting the emergency supervision requirements at special times.

2.2.2 Septic Tank

Septic tank, is the simplest sewage treatment device with the lowest investment, with its concept dating back to the 19th century. Mouras from France designed a new type of septic tank in which both the inlet and outlet pipes were deeply inserted under the liquid surface to form a water seal (Zhang et al., 2021) in 1860. "Mouras pond" is the origin of modern septic tanks and is also seen as the beginning of anaerobic biological treatment technology (Fan et al., 2017). Cameron and Cummins in England named the modified "Mouras tank" as "Septic tank" and patented it in 1895. A two-grid "Imhoff tank" was proposed by Imhoff from German based on the septic tank in 1905 to strengthen the separation of solid matter and effluents, and has been extensively used in the primary treatment of urban drainage worldwide.

Static separation and anaerobic fermentation are the treatment principles of septic tanks (Kamel & Hgazy, 2006). The manure is formed into the scum layer, the liquid layer, and the sediment layer under the combined action of gravity and buoyancy after entering the septic tank, as shown in Fig. 2.3. Preliminary treatment of sewage can be achieved through decomposing organic matters into methane, carbon dioxide, hydrogen sulfide, and ammonia via anaerobic bacteria in the tank during the long retention time. Meanwhile, high ammonia nitrogen and high pH environment in the anaerobic digestion process can kill pathogens such as parasite eggs, realizing the harmless treatment of effluent and slag (Fan et al., 2017). After sufficient stabilization, cleaned solids can be used as fertilizers, and the liquid in the middle layer can be used for farmland irrigation or directly discharged under low environmental requirements, otherwise, further treatment is needed before discharging.

The three-grid septic tank (Fig. 2.4) widely used today was developed in China, with one grid added for storing the cured manure. In this way, the necessary hydraulic

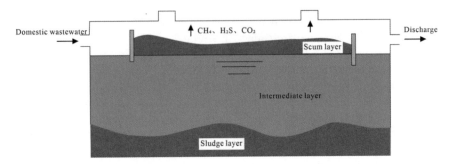

Fig. 2.3 Septic tank

retention time is not affected while feces can be easily taken out (Zhang et al., 2021). The three-grid septic tank featuring a simple structure and satisfactory treatment effect has the same treatment principle as the single-grid septic tank. Feces are precipitated in the first grid and degraded by fermentation, the fecal liquid is further fermented upon flowing into the second grid, and harmless fecal fluid is stored in the third grid (Cheng et al., 2018; Zheng et al., 2022).

The 3:1:1 structure ratio proposed in *"Code for Design of Water Supply and Drainage in Buildings"* (GB50015-2019), the 2:1:1 structure ratio proposed in

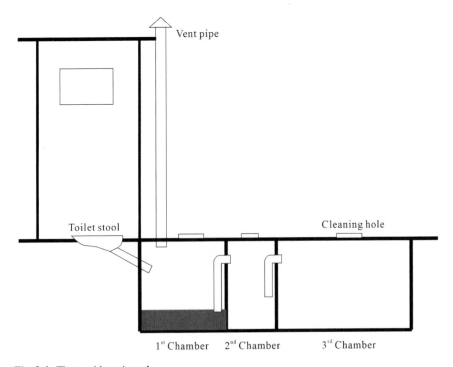

Fig. 2.4 Three grid septic tank

"Technical Specifications for Town and Village Drainage Engineering" (CJJ 124-2008), and the 2:1:3 structure ratio proposed in *"Rural Household Toilet Hygiene Standard"* (GB 19379-2012) are three mainstream structures of three-grid septic tanks. Among them, the 3:1:1 type septic tank with a retention time of 12–24 h is used for removing suspended solids in domestic sewage. It serves as the primary transitional domestic sewage treatment structure before the discharge into the urban downcorner network. The 2:1:1 type septic tank with a retention time of 24–36 h, has the same function as that of the 3:1:1 structure, and the 2:1:3 septic tank with a retention time of no less than 60 days is suitable for the harmless treatment of feces in rural household toilets.

Liner should be performed on B&Rck or reinforced concrete septic tanks as well as prefaB&Rcated septic tanks using glass fiber reinforced plastics or polyethylene to prevent the pollution of groundwater and the surrounding environment. Meanwhile, odor-resistant and explosion-proof measures should be also taken. The septic tank is more applicable for treating toilet sewage. Domestic grey water should not be discharged into the septic tank, otherwise, its treatment effect under the design retention time cannot be guaranteed, and the feces diluted by the domestic miscellaneous drainage may have less value of fertilizers after treatment (Zheng et al., 2022). Moreover, the septic tank should be cleaned in time at an interval of 3–12 months.

Featuring a simple structure, simple operation, energy-saving, and low investment cost, septic tank is superior to other domestic sewage treatment processes. But its limited treatment capacity should be also concerned. Its discharge should be first considered as resource utilization. If not, further treatment should be performed before discharge to reach the discharge standard.

2.2.3 Biofilm Process

The biofilm process is to remove pollutants in sewage using the dense biofilm attached to the carrier surface, a sewage treatment technology juxtaposed with the activated sludge process. Biofilms are formed by microorganisms wrapped by extracellular macromolecular polymers (EPS). EPS is composed of polysaccharides, proteins, and other matters, which can immobilize microorganisms on the carrier like glue to enable their growth and reproduction (Zhang, 2015). Organic pollutants of sewage enter the biofilm through diffusion and are extracted and decomposed by microorganisms as nutrients. Then, the sewage is purified. The multiplication of microorganisms, the growth, thickening, and shedding of biofilms are coupled with the process.

The biofilm process is applicable for rural domestic sewage treatment with small and medium water volumes as it has characteristics that the activated sludge process does not have, such as strong adaptability to changes in water quality and quantity, convenient management, long sludge age, diversified species, and few excess sludges.

The biofilm process frequently used in rural sewage treatment is composed of an anaerobic biofilm tank, biological filter, biological contact oxidation tank, and rotating biological disc.

i. **Anaerobic Biofilm Tank**

Limited by economic conditions in rural areas, the anaerobic sewage treatment technology with no energy consumed is more advantageous than the anoxic technology. Moreover, the biofilm process can better meet the actual needs of rural domestic sewage treatment with its features of easy control, strong shock load resistance, difficulty in biomass loss, and less dependence on the control of process operating conditions (Cui et al., 2021). The anaerobic biofilm tank is suitable for the treatment of rural domestic sewage as it combines anaerobic technology and biofilm process.

The anaerobic biofilm tank is an anaerobic reactor with carriers. Anaerobic microorganisms grow on the surface of the carrier in the form of biofilm. When the sewage flows through, its organic matters are absorbed by the biofilm and removed under the combined action of biofilm adsorption, microbial metabolism, and carrier interception. Biochemical reaction in the anaerobic biofilm tank is an anaerobic digestion process that degrades organic matter into gases (primarily methane and carbon dioxide). This process can be divided into the hydrolysis stage (microorganism degrading the high molecular organic matter into low molecular soluble organic matter via extracellular enzyme), acidification stage (low molecular organic matter being decomposed into volatile fatty acids, ethanol, and lactic acid, etc.), acetogenic stage (product in the acidification stage being further decomposed into acetic acid, H_2, CO_2, etc.), and the methanogenic stage (generating methane and new cells) (Henze et al., 2008). This technology can effectively remove organic matters in sewage, and explosion-proof measures should be taken.

The selection of carrier is essential for the anaerobic biofilm tank technology, which has a direct impact on the treatment effect of sewage (Zhang, 2021). Hence, the carrier should be selected with consideration of its surface properties, biochemical stability, roughness, biotoxicity, and economical applicability. Ceramics, zeolites, plastics, and activated carbon are commonly used carriers in practical application engineering (Karadag et al., 2015). The anaerobic biofilm tank technology is employed in the anaerobic tank part of the Japanese combined treatment type Johkasou with commonly used carriers of plastic materials such as polyethylene (PE) and polypropylene (PP) in a variety of shapes, such as grid-based cylinder, flat, grid-based flat, grid-based cylinder, and sphere, as shown in Table 2.4.

The SS removal effects of various shapes and materials of carriers are compared in a combined treatment Johkasou. The findings showed that grid-based plate (the first anaerobic tank) and spherical (the second anaerobic tank) carrier are the carrier combination with the highest SS removal rate (Ogawa & Iwahori, 2002).

The reaction rate and sludge production rate of the anaerobic treatment process are far lower than that of the oxic treatment process, removing up to 90% of the excess sludge production (Henze et al., 2008). The hydraulic retention time of the anaerobic biofilm tank, in general, is taken as two days to five days in practical engineering applications with the sludge discharge interval of three months to 12 months.

To sum up, the low removal rate of TN and TP of the anaerobic biofilm tank should be noted although it has many advantages. Apart from being used for farmland irrigation, the discharge from the treatment plant can be further treated for the removal

Table 2.4 Common anaerobic carriers in Johkasou

Shapes	Reference picture
Grid-based flat shape	
Sphere	
Grid-based cylinder	
Flat shape in the form of loofah fiber	

of TN and TP with other processes. Moreover, the long biofilm domestication time of the anaerobic biofilm technology and the difficulty in the control of old biofilms shedding are critical for its application (Escudié et al., 2011).

ii. **Biological Filter**

Biological filter is a typical biofilm technology developed from the practice of sewage irrigation according to the principle of soil self-purification (Liu, 2019). It is one of the earliest and most widely applied processes in the field of sewage treatment, with a history of more than 100 years. It can be divided into various process types such as common biological filter (trickling filter), high-load biological filter, tower biological filter, and aerated biological filter. Common biological filters are most frequently used in rural sewage treatment.

Fig. 2.5 Basic process flow of biological filter

The biological filter is filled with fixed bed carriers. The sewage flows through the carrier layer from top to bottom in the form of dropwise spraying and is purified by contacting with the biofilm on the carriers. It uses natural ventilation for oxygen supply, and is characterized by simple operation, low operating cost, low sludge production, and stable operation. Its basic process flow is shown in Fig. 2.5. The sewage enters the biological filter for treatment upon the removal of suspended solids and other pollutants that may block the carrier, and then passes through the secondary secondary sedimentation tank to retain the biofilm that falls off the biological filter, ensuring water quality. In addition, a equalization tank should be set when the water quality and quantity fluctuate greatly.

The common biological filter is comprised of the filter, carriers, water distribution device, and drainage system. To be specific, plastics and gravels such as polyethylene, polystyrene, and polyamide are common carriers, and materials with large specific surface area and high porosity such as corrugated plates, porous screened plates, and plastic honeycombs are plastic carriers. Either fixed or mobile water distribution device can be adopted to evenly distribute sewage on the surface of the filter tank. The drainage system, consisting of a seepage device, water collection ditch, and general drainage ditch, is set at the bottom of the filter for drainage and ventilation. The hydraulic load of the ordinary biological filter should be ranged from 0.1 to 0.5 m^3/m^2/h (Administration and for Market Regulation, 2019).

The biological filter is applicable for centralized treatment of rural domestic sewage due to its characteristics of simple operation and low operating cost. But it also faces problems such as low processing load, large coverage, susceptible to plugging, backwash, large water head loss, and filter flies attracted by stinking (Yue et al., 2013).

iii. **Biological Contact Oxidation Tank**

First proposed by Wring at the end of the nineteenth century (Wang, 2015), the concept of biological contact oxidation was developed on the basis of biological filters and has grown into a proven aerobic biofilm sewage treatment process widely used in the treatment of rural sewage, urban sewage, and industrial sewage.

The carrier of the biological contact oxidation tank is completely immersed in the sewage and should be supplied with oxygen through air blow and aeration, so it is also known as the "submerged biological filter" or "contact aeration process". The biological contact oxidation system consists of fixed bed carriers and supports, an aeration system, a water inlet and outlet device, a sludge discharge pipeline and a tank. The sewage is fully contacted with the biofilm fixed on the surface of the carrier under aerobic conditions. Organic matters and nutritive salts in sewage can be removed through microbial degradation. After that, the sewage is purified. Sufficient dissolved oxygen in the system, abundant species of microorganisms, and old biofilm shedding by the aeration are beneficial to maintaining the biofilm activity. The technology is characterized by high biomass (with the biofilm amount reaching 8000–14,000 mg VSS/L), strong removal ability of organic matters, strong adaptability to impact load, less sludge production, simple operation, and easy to manage (Zhang, 2015).

Carriers such as suspended carriers and fixed carriers can be used in the contact oxidation sewage treatment (HJ 2009–2011) with flexible settings, as shown in Fig. 2.6. Common carrier materials are plastics, glass fiber reinforced plastics, and fiber in the shapes of honeycomb, cylinder, corrugated plate, and bundle. The contact oxidation process includes the primary-stage contact oxidation process and multi-stage contact oxidation process, and the secondary-stage or multi-stage contact oxidation process can be used for organic matters of high concentration, as shown in Fig. 2.7. The BOD$_5$ of the biological contact oxidation tank is ranged from 0.15 to 0.18 kg/m^3/d when the treatment capacity is less than 5 m^3/d, and should be ranged from 0.15 to 0.2 kg/m^3/d when the treatment capacity is more than 5 m^3/d according to GB/T51347-2019 (Administration and for Market Regulation, 2019).

The biological contact oxidation tank is used for household sewage treatment and centralized treatment of rural domestic sewage. The combined process of anaerobic tank and aerobic tank should be used when denitrification is required. With a poor effect in biological removing of phosphorus, chemical phosphorus removal is normally required in response to the requirements of phosphorus removal. Note that the treatment effect can be affected by carrier blockage, and uneven water and gas distribution under improper design or operation.

iv. **Rotating Biological Disc Reactor**

The rotating biological disc reactor, born in Germany in the 1920s (Chan & Stenstrom, 1979), is a typical biofilm treatment technology developed from biological filters, and has been widely used in the treatments of rural sewage, urban sewage, and industrial sewage.

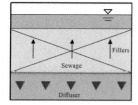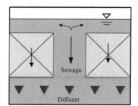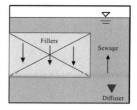

Fig. 2.6 Carriers setting process (left: full aeration; middle: central aeration; right: side aeration) (Matsuo, 2015)

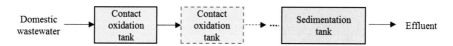

Fig. 2.7 Flow chart of contact oxidation process

Consisting of a disc, a contact reaction tank, a rotating shaft, and a driving device, it performs sewage treatment based on the rotation of the disc covered with biofilm. In general, multiple sets of discs are strung on the rotating shaft, with 35–45% of the rotating disc area immersed in sewage. With the rotation of the shaft and the disc, the biofilm on the disc contacts with organic matters, ammonia nitrogen, and other pollutants while being rotated to the sewage, and contacts with oxygen while being rotated to the air. In circulation, pollutants are discomposed by microorganisms to purify the sewage (Chan & Stenstrom, 1979). The matured biofilm is peeled under the action of the shear force of water flow and sedimented in the secondary secondary sedimentation tank, as shown in Figs. 2.8 and 2.9.

The rotating biological disc reactor has the advantages of high efficiency, simple structure, easy to main, and low power consumption, together with a strong resistance to water shock load (treat the sewage with a BOD_5 value of 10–10,000 mg/L) and high microbial concentration (reaching up to 50,000–60,000 mg VSS/L), multiple species of microorganisms, long sludge age, low sludge production rate (accounting for roughly 1/2 of the activated sludge process), no need of aeration and sludge return (Gao and Li, 2018).

Discs in this process are mostly circular and regular polygonal mesh plates or corrugated plates made of polyethylene or polyester fiberglass, featuring light material, high strength, corrosion resistance, and large specific surface area. In general, the disc diameter is ranged from 2 to 4 m with the disc space between 10 and 35 mm and the rotational linear velocity ranging from 15 to 18 m/min (Zhang, 2015). Uniaxial multi-stage rotating biological discs should be used for centralized sewage treatment

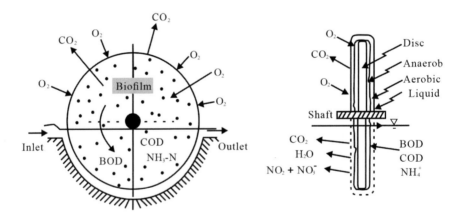

Fig. 2.8 Sewage treatment mode of rotating biological disc reactor (Zhang, 2015)

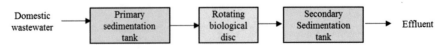

Fig. 2.9 Process flow of rotating biological disc reactor

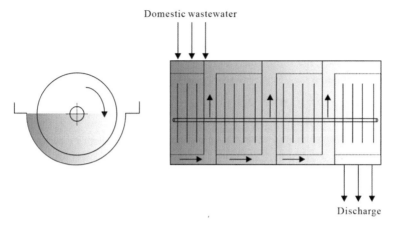

Fig. 2.10 Single-axis multi-stage rotating biological disc reactor

in rural areas with no less than three stages, as shown in Fig. 2.10 (Administration and for Market Regulation, 2019).

The rotating biological disc reactor can be used for the centralized treatment of rural domestic sewage. Note that there is a risk of reduced pH value in the contact reaction tank with the nitrification reaction. Moreover, the tiny SS that is difficult to remove may lead to low visibility of the treated water, causing problems such as odor and breeding of flies and mosquitoes (Matsuo, 2005).

2.2.4 Activated Sludge Process

The activated sludge process developed based on the principle of water self-purification has a history of more than 100 years and is a biological water treatment technology most extensively used today. It can purify the sewage using the free-flowing tiny "sludge particles" (a flocculating constituent formed by a large number of microorganisms and EPS) suspended in the sewage, which is different from the biofilm process that depends on the biofilm attached to the carrier for purification. The activated sludge flocculating constituent flows freely in the sewage and adsorbs organic matters with the action of aeration or mechanical stirring. The microorganisms in the flocs carry out aerobic metabolism with oxygen in the oxic environment, through which, organic matters are decomposed into carbon dioxide (CO_2) and water (H_2O) to purify sewage. During this, microorganisms obtain nutrients and energy for multiplication and are finally removed in the secondary sedimentation tank via solid–liquid separation. As the activated sludge can be mobilized freely in sewage, it has a higher chance of contacting pollutants and oxygen. In that case, a higher pollutant removal efficiency with a more flexible process can be realized compared with the biofilm process.

The activated sludge process has experienced continued innovation and development in terms of theory, technology, and process, and the derived technologies can efficiently degrade organic matter and remove nitrogen and phosphorus using microorganisms at the same time. Currently, the traditional activated sludge process and its modified process have been widely used in rural sewage treatment.

i. Traditional Activated Sludge Process

B&Rtish scientists Ardern and Lockett announced the birth of the activated sludge process at the B&Rtish Chemical Society and defined its basic principle that the solid suspended matter generated in sewage should be recycled and returned to sewage for accumulation for its purifying effect upon the introduction of air into the sewage, rather than being removed (Henze et al., 2008).

Upon the introduction of air into the sewage (aeration), organisms such as bacteria, protozoa, and metazoans can be multiplied with oxygen and organic matters in sewage and aggregated to form gelatin yellow–brown flocculating constituent or activated sludge. The suspended activated sludge will be sedimented rapidly as soon as the aeration is stopped, with clear water obtained upon the separation of muddy water. Note that the sludge in the anaerobic treatment process cannot be called activated sludge.

From the perspective of microbial engineering, the activated sludge process refers to the rapid adsorption of colloidal and granular macromolecular organic matters dissolved in sewage upon contact with activated sludge and converting organic matter into small molecules finally taken into cells for metabolism by extracellular enzymes of microorganisms (Matsuo, 2005). Specifically, part of organic matter (roughly 50%) is metabolized to be newly-synthesized bacterial cells while the other is catabolized to generate CO_2, H_2O, and energy used to maintain life activities and anabolism of microorganisms, as shown in Fig. 2.11.

Figure 2.12 shows the traditional activated sludge process flow with a secondary sedimentation tank, oxygen supply device, and reflux equipment. After removing larger particles of suspended solids by the primary secondary sedimentation tank,

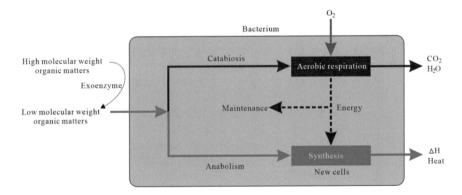

Fig. 2.11 Metabolic reaction of heterodoxic microorganisms

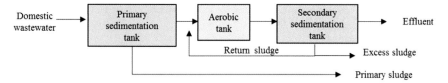

Fig. 2.12 Process flow of traditional activated sludge process

the sewage is mixed with the reflux sludge flowing out from the bottom of the secondary sedimentation tank. Then the sewage flows into the aerobic tank which provides the oxygen required by microorganisms for aerobic metabolism, and plays the role of suspending the activated sludge to make it fully mixed with the sewage to improve the efficiency of pollutant removal. At last, mud and water are separated in the secondary sedimentation tank. The clear water in the upper layer is discharged as treated water, and part of the sedimented sludge is mixed with the new influx sewage before returning to the aerobic tank as return sludge, while the rest is discharged from the equipment for disposal as excess sludge. Then, the whole process of sewage treatment is done. The volume load of the activated sludge process is set at 0.1 kg BOD_5 (m^3/d) in rural sewage treatment.

ii. A/O Activated Sludge Process

The activated sludge process has become a mainstream process of sewage treatment with its features of high efficiency, flexible operation mode, and low daily operating costs. Based on this, a variety of variant processes has been evolved to remove nitrogen (N) and phosphorus (P) elements in sewage, so as to prevent the eutrophication of received water. Specifically, the anoxic/oxic activated sludge process (A/O) is a typical biological denitrification process.

Nitrogen in sewage mainly exists in the form of organic nitrogen (protein, amino acid, etc.) and NH_3–N, which originated from the human body's metabolism of ingested protein. Biological denitrification achieves the purpose of denitrification by converting organic nitrogen and NH_3–N into nitrogen gas (N_2) based on some obligate bacteria (ammonifying bacteria, nitrifying bacteria, and denitrifying bacteria) (Henze et al., 2008). Among them, organic nitrogen is decomposed into NH_3–N via the effect of ammonifying bacteria. After that, NH_3–N is oxidized to nitrite (NO_2^-) by nitrite bacteria under the aerobic state (Table 2.5), and then further oxidized to nitrate (NO_3^-) by nitrate bacteria. These two reaction processes are nitrification reactions, and nitrite bacteria and nitrate bacteria are nitrifying bacteria. At last, NO_3^- is reduced to N_2 by denitrifying bacteria using organic matters (electron donor) in the anoxic environment (Table 2.5) before discharged into the air (denitrification).

The above nitrification and denitrification processes, together with nitrogen fixation, constitute the traditional nitrogen cycle process, NO_3^- can be excluded from the circulation to shorten the reaction process and save the input of oxygen and organic matters from the point of the technical process. Some microorganisms can generate nitrous oxide (N_2O) under micro-aerobic conditions using NO_3^- and NH_2OH and

Table 2.5 Definitions of anaerobic, anoxic, and oxic environments and their biochemical reactions

	Anaerobic (A)	Anoxic (A)	Oxic (O)
Definitions	No molecular oxygen (O_2), no combined oxygen (NO_2^-, NO_3^-) 23	No molecular oxygen, with combined oxygen	With molecular oxygen and combined oxygen
Definition of operation and maintenance perspectives	Dissolved oxygen (DO) $\leq$ 0.2 mg/L	0.2 mg/L < DO $\leq$ 0.5 mg/L	DO $\geq$ 2 mg/L
Major biochemical reactions	Removal of some organic matters; phosphorus release of phosphorus-accumulating bacteria	Removal of some organic matters; denitrification reaction, removal of TN	Removal of organic matters; nitrification reaction and removal of ammonia nitrogen; phosphorus absorption of phosphorus-accumulating bacteria

convert NO_2 into NO gas under anoxic conditions, according to studies. Nitrogen can be removed from the liquid phase using these two processes. But the toxicity of NO and the strong greenhouse effect of NO_2 (approximately 310 times that of CO_2) should be highly concerned. By contrast, Anammox bacteria can generate N_2 using NH_4^+ and NO_2^- under anaerobic conditions, which is considered the next-generation biological denitrification technology as it can shorten the nitrogen cycle without generating harmful by-products (Henze et al., 2008).

The traditional nitrogen cycling route is still the mainstream biological denitrification process. More precisely, A/O activated sludge process (Fig. 2.13) is constituted by adding an anoxic pond (A tank) and a nitrification solution return system based on the O tank of the traditional activated sludge process. The nitrifying liquid reflux system functions as refluxing the NO_3^- in the outlet of the aerobic tank to the anaerobic tank for denitrification. The anaerobic tank can also be set behind the aerobic tank. But setting the anaerobic tank in the front with the nitrification reflux (Fig. 2.13) can fully utilize organic matters in the raw water for denitrification reaction since the concentration of organic matters in the outlet of the aerobic tank is low, and the participation of organic matters is required in the denitrification reaction. Moreover, the intermittent aeration operation mode can be also adopted in the traditional activated sludge process for denitrification in the rural sewage treatment since the oxygen content (dissolved oxygen) in the oxic pond is reduced to reach the anoxic state after the aeration stops, as shown in Table 2.5. By doing so, it can provide conditions for the denitrification of denitrifying bacteria.

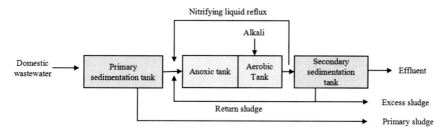

Fig. 2.13 A/O process flow

iii. **Enhanced Biological Phosphorus Removal (EBPR) and A²/O Activated Sludge Process**

Phosphorus, as an essential inorganic element of the vital activity of the organism, plays an important role in cell composition, metabolic reaction, and material transformation. The phosphorus in domestic sewage mostly come from urine and can be removed by the chemical process. In other words, phosphorus can be removed by the chemical precipitation of external chemical substances and phosphorus. But biological phosphorus removal remains the most economical and effective process for phosphorus removal.

As phosphorus is the inorganic element with the largest demand for microorganisms, 15–25% of phosphorus can be removed from domestic sewage via the anabolism of microorganisms in the traditional activated sludge process. The quantity of phosphorus absorbed by PAOs in the system can exceed the amount required for normal microbial anabolism, thereby achieving enhanced biological phosphorus removal (EBPR), as shown in Fig. 2.14. The phosphorus content in PAOs can reach as high as roughly 0.38 mgP/mgVSS, and the overall phosphorus content in the sludge in the EBPR system can reach approximately 0.06–0.15 mgP/mgVSS (Henze et al., 2008), far higher than 0.02–0.03 mgP/mgVSS of traditional activated sludge process (Matsuo, 2005).

The environmental condition of first being anaerobic (Table 2.5) and then oxic is required for concentrating phosphorus with PAOs. The metabolic process is shown in Fig. 2.15. PAOs ingest organic matters (mainly volatile fatty acids such as acetic acid) using the energy (ATP) generated by the hydrolysis of Poly-P, and finally store them in cells in the form of PHA in the anaerobic environment. The tricarboxylic acid (TCA) cycle or the degradation of glycogen (Mino et al., 1998) contributes to the reducing

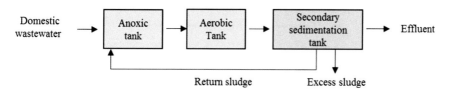

Fig. 2.14 Process flow of EBPR system

power required for the synthesis of Poly-P as a reduction product. Specifically, the former is known as the Comeau-Wentzel model (Comeau et al., 1986), while the latter is known as the Mino model (Mino & Matsuo, 1984). Besides, the release of PO_4^{3-} is accompanied by the hydrolysis of Poly-P, as shown in Figs. 2.15 and 2.16.

The PHA stored in the PAOs cells during anaerobism is used for catabolism and anabolism after entering the aerobic environment. The energy obtained is adopted for the synthesis of Poly-P or glycogen, so as to prepare for entering the next anaerobic environment. Specifically, PO_4^{3-} should be obtained for the synthesis of Poly-P, as shown in Figs. 2.15 and 2.16. PAOs can metabolize normally even if there is no intake of organic matter in an oxic environment since it stores PHA in an anaerobic environment. In this way, dominant bacteria will be formed by PAOs in the anaerobic-oxic environment.

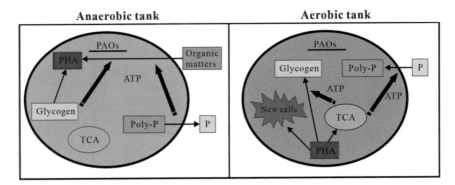

Fig. 2.15 Metabolism of PAOs

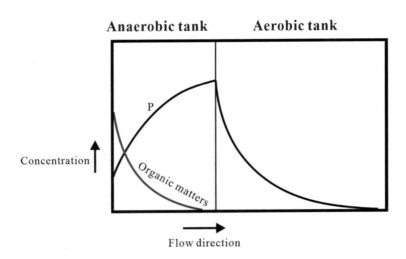

Fig. 2.16 Dynamics of organic matters and phosphorus in EBPR system

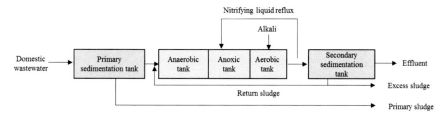

Fig. 2.17   A²/O activated sludge process flow

The anaerobic/anoxic/oxic activated sludge process (A²/O) can be obtained by adding a first-stage anaerobic tank (A) to the A/O process, as shown in Fig. 2.17. The P in sewage is absorbed by PAOs in the sludge and enriched in the sludge after the activated sludge flows through the anaerobic and oxic ponds in turn. Finally, the separation of muddy water and the discharge of excess sludge are performed to remove P, achieving the purpose of simultaneous denitrification and phosphorus removal.

iv.   **Oxidation Ditch**

As a kind of activated sludge process of extended aeration, oxidation ditch is also known as the continuous circulation reactor, which was originally a process for treating dairy farmer's sewage on a simple facility that emerged in the Dutch countryside and then formally put into engineering application in 1954 (Thakre et al., 2019). It is widely used in the treatment of rural and municipal sewage nowadays. Obral oxidation ditch, Carrousel oxidation ditch, and single-ditch oxidation ditch are the widely used oxidation ditches, and the Carrousel oxidation ditch with the longest history is the most widely used one worldwide.

The simplest continuous oxidation ditch process is shown in Fig. 2.18. The annular reaction tank is equipped with a secondary secondary sedimentation tank, and it has developed into a reinforced concrete structure from the originally simple excavated ditch. The mechanical aeration device is leveraged for oxygen supply, which can provide the required oxygen for activated sludge treatment and mix the activated sludge and sewage, driving the circulation of the mixture at a certain flow rate, and forming a unique flow pattern between the complete mixture and plug flow. The activated sludge experience aerobic and anoxic states during the circulation in the annular ditch, and organic matters and TN are removed at the same time.

An oxidation ditch can be seen as a completely-mixed aeration tank. Raw water has minor effect on the concentration of pollutants in the ditch, as it can be diluted by dozens or hundreds of times of flow upon entering the tank. The ditch is normally operated at a low sludge load of 0.03–0.10 kgBOD$_5$ (kgMLSS/d) (Ai and Cui, 2018; Administration and for Market Regulation, 2019) with the ability to resist the impact load of water quality and quantity. Moreover, a certain processing capacity can be maintained at a low temperature (roughly 5 °C). The nitrification reaction can be performed easily with the oxidation ditch technology for its small sludge load and long retention time. Hence, the continuous water intake and intermittent aeration

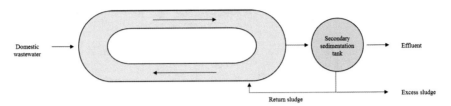

Fig. 2.18 Process flow chart of single-ditch oxidation ditch

operation mode can be adopted in the single-ditch oxidation ditch for denitrification, and the removal rate of TN can reach roughly 70% by setting the anoxic area in the annular waterway.

To sum up, the traditional activated sludge process, A/O activated sludge process, A²/O activated sludge process, and oxidation ditch are all applicable for centralized treatment of rural domestic sewage. By comparison, traditional activated sludge process, A/O activated sludge process, and A²/O activated sludge process, with higher requirements for operation, are called "operation-dependent" processes (Wang, 2018), frequently requiring special persons for process inspection or maintenance. Hence, they face great difficulty in implementation in rural areas. Moreover, there are also problems such as high energy consumption (blowers, reflux systems, etc.), poor resistance to shock loads, susceptibility to sludge bulking, and unstable biological phosphorus removal (for instance, after mixed with rainwater) (Yoshida et al., 2005). The oxidation ditch process featuring simple treatment flow, stable and reliable process, simple operation and maintenance, and low investment, can be used for the centralized treatment of rural domestic sewage. Note that the oxidation ditch requires a large land area for its long retention time and shallow tank.

2.2.5 Membrane Bioreactor

Membrane bioreactor (MBR) is a sewage treatment process organically combined by membrane separation and biological treatment. The concept of MBR originated in the United States. Dorr-Oliver built the world's first MBR process sewage treatment plant (14 m³/d) in the 1960s. MBR experienced rapid growth with the development of material technologies and the reduction of energy consumption costs in the 21st century after a long R&D development stage (Hao et al., 2018). The technology is extensively applied in various fields, such as domestic sewage treatment and industrial sewage treatment (Fig. 2.19).

MBR has replaced the secondary sedimentation tank of the traditional activated sludge process by introducing membrane separation technology. While retaining the advantages of the activated sludge process, it has greatly improved the solid–liquid separation efficiency, reduced land use, and improved the quality of effluents. The effluents can be directly used for compensating landscape water and recycling

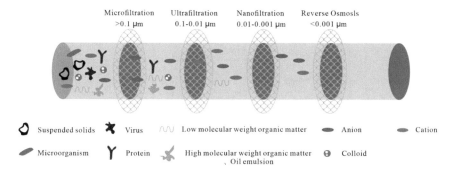

Fig. 2.19 Separation range of different membranes (Patsios & Karabelas, 2010)

reclaimed water (Henze et al., 2008). Microfiltration membranes and ultrafiltration membranes are adopted in MBR membrane modules in various forms such as hollow fiber, flat and tubular types. There are also various types of membrane materials, including PVDF, PP, PVC, PE, PES, and ceramic membranes (Zhang et al., 2020). MBR is comprised of a bioreactor, membrane module, and smallwater pump (Henze et al., 2008). MBR can be classified into immersion membrane bio-reactor (IMBR) and separate membrane bio-reactor (SMBR) by processing technology, as shown in Fig. 2.20. IMBR consumes less energy and has broader applications than SMBR (Lin, 2015). Furthermore, MBR can be classified into anaerobic MBR, facultative MBR, and aerobic MBR processes according to the DO concentration level in the reaction tank.

MBR is superior to the traditional sewage treatment process in the following aspects. First, a compact structure (small floor space). As no secondary sedimentation tank is needed, it can decrease 1/5 to 1/3 of land use in the traditional activated sludge process. Second, high quality of effluents. The membrane module can efficiently intercept SS, turbidity, bacteria, and viruses. Third, the MLSS can reach up to 10–20 g/L with a large volume load, strong resistance to the impact of water quality and quantity, and less output of excess sludge. Fourth, HRT and SRT of the reactor are completely separated, and the HRT can be adjusted according to the treatment effect, with more flexible operation control. Long SRT can enrich the long-generation microorganisms such as nitrifying bacteria to promote the removal of ammonia nitrogen and refractory organic matters. Finally, there is no risk of sludge bulking (Jia, 2016). The compact structure (that is, small coverage) and good outlet quality are the absolute advantages of MBR.

Prominent disadvantages such as high operating costs and membrane pollution can be found in MBR. Expensive membrane modules, various accessory facilities, and high energy consumption are the causes of high operating costs. Relevant studies revealed that the investment cost of MBR is between 2000 and 5000 yuan/m³/ d (roughly 3800 yuan/m³/d on average), 10–30% higher than the traditional activated sludge process plus three-stage filtration combined process (Krzeminski et al.,

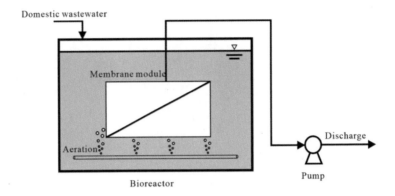

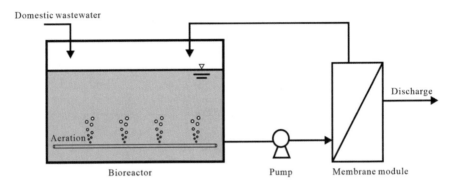

Fig. 2.20 IMBR (upper) and SMBR (lower)

2012). Regarding energy consumption, maintaining membrane flux via pressurization (membrane pollution leads to a decrease in flux) and higher aeration rate (the aeration rate should be increased to relieve the membrane pollution) cause the energy consumption of MBR as high as roughly 2 kW h/m^3, 60–900% higher than the traditional activated sludge process (approximately 0.3 kW h/m^3) (Sun et al., 2016). In addition, although remarkable scientific research results and technologies have been made to address the problem of membrane pollution as a hot spot of MBR research, the problem is not solved fundamentally (Meng et al., 2009). Online cleaning for maintenance (once in 1–3 months) and offline chemical cleaning for restoration (once in half a year to 1 year) are required, resulting in the reduced service life of the membrane and increased operating costs. What's more, the high degree of automated operation poses higher requirements for the operation and maintenance personnel.

To sum up, MBR is a process with outstanding advantages and disadvantages and is applicable for domestic sewage treatment in rural areas where land space is strictly limited with high discharge standards, a need for reclaimed water reuse, and economic conditions allowed.

2.2.6 Natural Biological Treatment

Natural biological treatment is a process technology that purifies sewage using natural ecological or artificial ecological functions. Sewage is purified through the physical and chemical effects of soil or artificial carriers, the purifying function of natural organisms, and the interception and absorption of aquatic plants. It has been widely applied with its features of high purification efficiency, low operation and maintenance costs, and low management level required. Constructed wetlands and stabilization ponds are two commonly used processes in rural sewage treatment.

i. Constructed Wetlands

Constructed wetland is an artificially designed and constructed complex consisting of substrates (fine sand, gravel, etc.), wetland plants, organisms, microorganisms, and water bodies by simulating the natural wetland (Li et al., 2018). It has been widely used in domestic sewage treatment since the world's first constructed wetland for sewage treatment was built more than 100 years ago in Yorkshire, England in 1903. Specifically, the constructed wetland can purify sewage through a series of filtration, adsorption, co-precipitation, ion exchange, plant absorption, and microbial decomposition combing physical, chemical, and biological treatment processes (Kadlec & Wallace, 2008). It can be subdivided into surface flow, horizontal subsurface flow, and vertical subsurface flow as per the characteristics of water flow, as shown in Fig. 2.21.

Surface flow constructed wetlands are open waters, with the appearance most approaching natural wetlands. They are planted with floating plants, emergent aquatic plants, or phreatophytes. The sewage flows horizontally in the surface layer with a shallow water level between 0.1 and 0.3 m, presenting a favorable oxygenation effect. Organic matters are removed by sedimentation, filtration, oxidation, reduction, and adsorption during the flow. The surface flow wetland is similar to the natural wetland in terms of structure, and can serve as the habitat of insects, fish, and birds, presenting satisfactory ecological benefits and landscape effects. But there are also problems such as low pollutant loads, susceptibility to mosquito breeding in summer, and significant impact by temperature (Wu, 2014).

The horizontal subsurface flow constructed wetland is the most popular constructed wetland sewage treatment system worldwide with aquatic plants as surface vegetation, surface soil and lower gravel as matrix carriers, and water propelling in horizontal infiltration from the inside of the matrix bed. Unlike the surface flow constructed wetland, the role of its matrix carriers is fully exerted. The flowing sewage is purified via the function of biofilm on the surface of the matrix and the interception and adsorption of the matrix and the huge plant root system. Besides, plant roots have favorable oxygen releasing capacity to form an oxic, anoxic, and anaerobic microenvironment near the root system, which is conducive to the removal of organic matters and TN, according to Kiehuth's "root zone theory" (Hu, 2011). The horizontal subsurface constructed wetland is characterized by high pollutant load, good thermal insulation, and no mosquitoes breeding as the sewage is not exposed

Fig. 2.21 Three types of
constructed wetland

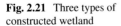

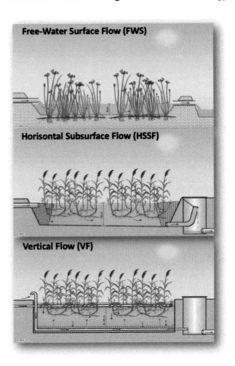

Fig. 2.21 Three types of constructed wetland

to the air. But it has the disadvantages of high investment as well as complicated construction and management.

The vertical flow constructed wetland is a process developed from the horizontal subsurface flow constructed wetland, which has the characteristics of both the horizontal subsurface flow and the surface flow constructed wetlands. Sewage is purified through passing by wetland plants and matrix carriers in vertical flow from top to bottom. The system adopts intermittent operation with oxygen accessing to the wetland through atmospheric diffusion and plant rhizome transportation, which has a better oxygen recovery than the horizontal subsurface constructed wetland. It has the strengths of high pollutant load and small floor space, but is defective in high investment, complex construction and management, and susceptible to plugging and mosquito breeding (Wang, 2019).

Constructed wetland technology also beautifies the landscape with a large buffer capacity and simple process. Note that constructed wetlands are normally used for advanced treatment, in other words, biological treatment technology should be adopted to reduce the concentration of pollutants before sewage enters the constructed wetland. As they usually require large land occupation, they are suitable for rural areas with rich land resources. The constructed wetland is designed according to the calculation of pollutant load and hydraulic load, as shown in Table 2.6, and desludging should be performed regularly during operation and maintenance.

Table 2.6 Main design parameters of constructed wetlands (Administration and for Market Regulation, 2019)

	Surface flow constructed wetland	Horizontal subsurface flow constructed wetland	Vertical subsurface flow constructed wetland
BOD_5 surface load/ $[g(m^2 \, d)^{-1}]$	≤ 4.5	≤ 10	≤ 20

ii. Stabilization Pond

Stabilizing pond, also known as oxidation pond or biological pond, is a biological treatment technology set with a causeway and anti-seepage layer to treat sewage using the natural biological purification function (Zhang, 2015). The first stabilization pond system was built in Texas, USA, in 1901 (Sopper & Kardos, 1973). With more than 100 years of proven practice, it can be effectively used for the treatment of domestic sewage. The purification process of the stabilization pond is similar to the self-purification process of the river. By flowing slowly after entering the stabilization pond, the sewage is diluted, precipitated, and purified by the combined effects of microorganisms, micro-fauna (protozoa, metazoa), algae, and aquatic plants in the long retention process (Cao et al., 2004). It can be divided into oxic stabilization pond, facultative stabilization pond, anaerobic stabilization pond, and aeration stabilization pond by the type of microbial dominant groups and the concentration of dissolved oxygen (Zhang, 2015).

Oxic ponds are shallow with roughly 0.5 m in depth, as shown in Fig. 2.22, which can be permeated by sunlight. The oxygen required for the metabolism of aerobic microorganisms in the pond is provided by the photosynthesis of algae and the atmospheric reoxygenation of the water surface. It can be seen that the system is essentially an algal–bacterial symbiotic system. More specifically, the algae in the pond release oxygen through photosynthesis under sunlight; oxic microorganisms produce CO_2 after oxic metabolism is performed on organic matters; and algae can absorb CO_2 and nutritive salts such as NH_4^+ and PO_4^{3-} in sewage (Han, 2011).

Facultative pond is the most commonly seen stabilization pond (Figs. 2.22 and 2.23). With a depth of 1–2 m, the pond has an oxic zone (the upper layer of the pond surface, with significant algae photosynthesis and sufficient dissolved oxygen), an anaerobic zone (at the pone bottom, anaerobic fermentation is performed on sediment by anaerobic microorganisms), and an in-between anoxic zone (Mahapatra et al., 2022). Pollutants such as organic matter, NH_4^+ and PO_4^{3-} are removed in the algal–bacterial symbiotic system in the oxic zone after the sewage flows into the pond. Denitrification is carried out in the anoxic zone. Refractory organic matters in the sewage and the dead algae are precipitated in the anaerobic zone at the pond bottom, which can be converted into organic acids by anaerobic fermentation and then partially diffused to the aerobic zone and anoxic zone for decomposing, while some organic acids are decomposed via anaerobism into methane (CH_4) and CO_2. In that case, the purification reaction in the facultative pond involves different aspects, and the biological phase in the system is also more abundant.

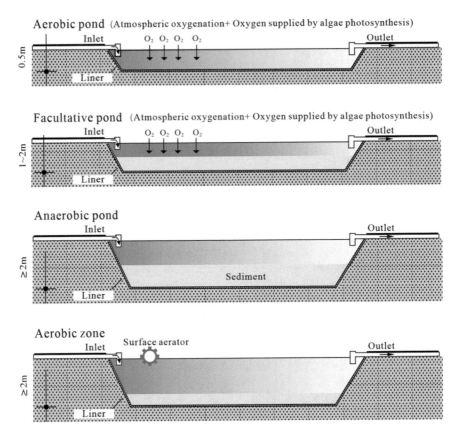

Fig. 2.22 Oxic stabilization pond, facultative stabilization pond, and anaerobic stabilization pond

The anaerobic pond is deep, over two meters in general, and is in an anaerobic state as a whole, according to Fig. 2.22. Its BOD_5 surface load is remarkably higher than that of aerobic and facultative ponds, reaching up to 20–40 g/(m^2 d). Anaerobic fermentation is carried out in the pond, leading to a slow purifying rate and long sewage retention time.

The aerobic zone is normally more than two meters in depth. Oxygen is provided by surface aerators, with high oxygen supply efficiency. But the growth and photosynthesis of algae are inhibited under the aeration condition. Compared with the anaerobic ponds with similar depths, the aerobic zone has lower organic loads, which is more conducive to the effective removal of NH_3–N due to sufficient oxygen.

The stabilization pond is suitable for rural sewage treatment with sufficient land resources with its advantages of simple engineering, low infrastructure investment, low operation and maintenance costs, simple control, effective removal of organic matters in sewage, removal of nutrients to a certain extent, and not requiring sludge treatment. But it should be noted that problems such as low organic load (Table 2.7),

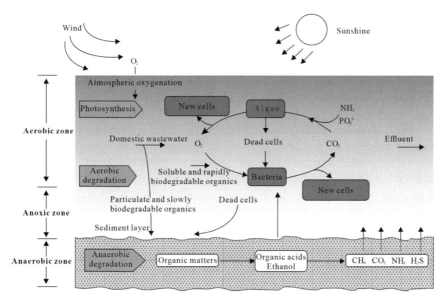

Fig. 2.23   Materials conversion in a facultative pond (Tchobanoglous & Schroeder, 1985)

Table 2.7   Main design parameters of stabilization pond (Administration and for Market Regulation, 2019)

Type of stabilization pond	Oxic pond	Facultative pond	Anaerobic pond	Aerobic zone
BOD$_5$ surface load (g/m^2/d)	1–3	3–10	20–40	5–40

large land use, poor environmental conditions, and the treatment effect susceptible to temperature and light can be observed. Hence, it is normally leveraged for advanced treatment of domestic sewage. Besides, liner is required at the bottom and surrounding of the stabilization pond to prevent groundwater pollution.

2.3   Sludge Treatment and Disposal in Rural Areas

2.3.1   Importance of Sludge Treatment and Disposal

The scum, sediment, and excess sludge produced by the primary secondary secondary sedimentation tank, secondary secondary sedimentation tank, and other process units are called sludge. As a by-product of sewage treatment, sludge is easily overlooked (Hang et al., 2004). Sludge treatment is dependent on sewage treatment. To be specific, sewage treatment is a process in which suspended, colloidal, and dissolved pollutants are converted into sludge by the adsorption and metabolism of microorganisms, and then precipitated from the liquid phase. Sewage treatment is merely an enrichment or transformation of pollutants in sewage. Closed-loop sewage treatment is not completed until the treatment and disposal of sludge are done. Therefore, sludge treatment should be seen as a part of sewage treatment.

There are large amounts of organic matter, nutrients, pathogenic bacteria, parasite eggs, heavy metals, and some toxic and harmful refractory organic substances in sludge besides microorganisms, showing high moisture content (above 99%), large volume, putrefactive, and generation of stink. They might easily cause secondary damage to groundwater and soil, posing threats to environmental safety and public health if not properly treated and disposed.

Moreover, approximately 55% of COD, 30–45% of N, and 85–95% of P are left in the sludge during sewage treatment (China Urban Water Supply and Drainage Association, 2021). CH_4, H_2, and other fuels with higher heat value can be obtained by rich organic matters via the anaerobic treatment. Also, resource utilization of energy and materials in disposal and treatment can be achieved using the stabilized product upon treatment as agricultural fertilizer (Dai, 2020).

Hence, attention should be given to the treatment and disposal of sludge in rural sewage treatment based on the attributes of "pollution" and "resources" of sludge.

2.3.2 Main Principles and Technical Options for Sludge Treatment and Disposal

Sludge treatment and disposal are two different concepts. Sludge treatment, in general, refers to the stabilized, reduced, and harmless treatment of sludge, including concentration (conditioning), dehydration, anaerobic digestion, oxic digestion, drying, and composting (aerobic fermentation). While, sludge disposal refers to the final consumption of sludge, including land use (agricultural use), landfill, and use and incineration of building materials.

Treatment and disposal of sludge require skilled technicians to operate the facilities, generating high construction and maintenance costs. Thus, the sludge treatment and disposal of rural sewage treatment should be performed in coordination with the urban sludge treatment and disposal system, otherwise a new sludge treatment plant can be built for centralized treatment. Furthermore, the treatment technology and process with low sludge yield should be prioritized in the treatment of rural domestic sewage in view of the scattered distribution of rural domestic sewage as well as the high cost of sludge collection and transfer.

The treatment and disposal of sludge in rural sewage treatment plants should be set under the principles of harmlessness, reduction, and stabilization, as well as the principle of resource utilization to achieve sustainable development of sewage treatment. Particularly, the sludge produced by rural sewage treatment is quite different from that produced by urban sludge, in terms of low content of heavy metals and great potential for land use (incl. agricultural use, landscaping, and soil improvement).

As economic development, treatment scale, residents' living habits, and natural climatic conditions are varied in the rural areas of B&R countries, sludge treatment and disposal technologies should be based on regional characteristics, comprehensive consideration of industrial structure, mud characteristics, processing scale, environmental conditions, economic factors such as the level of economic and social development and the final disposal process, so as to achieve the purpose of protecting the overall environment at a high level.

In general, gravity concentration and natural drying should be preferred for sludge concentration and dehydration, respectively, in the sludge treatment in rural areas. Farmland use or land use should be prioritized at the final disposal of sludge. Also, anaerobic digestion or oxic fermentation (composting) should be employed for stabilization and harmless treatment when the sludge is treated for landscaping and farmland utilization.

Since sludge and its leachate contain high concentrations of pollutants and pathogenic bacteria, spillage, leakage, and seepage should be avoided during storage and transportation. Otherwise, it will cause harm to the surrounding environment, surface water, groundwater, and soil. Meanwhile, the leachate should be also properly handled.

2.4 Technical Selections for Rural Sewage Treatment in B&R Countries

2.4.1 Main Principles for Selecting Rural Sewage Treatment Process

China has signed over 200 cooperation agreements under B&R with 147 countries and 32 international organizations since the initiative was proposed in 2014. With different climates, landforms, cultures, living habits, economic conditions, and the quality of rural domestic sewage in different B&R countries, the following principles should be followed in their rural sewage treatment:

i. Select the simplest process with strong resistance to shock load and stable compliance. The daily sewage volume in rural areas is small, and the water volume is mainly distributed in the morning, at noon, and in the evening, with a daily variation coefficient of 5–10 (Wang, 2018). Processes with strong anti-shock load capacity should be prioritized in the treatment of rural sewage, according to the characteristics of large changes in water quantity and quality.

ii. Select an energy-saving process with no power or less power consumption and low operating cost. The high operating costs of rural sewage treatment plants, small in scale and high in energy consumption per unit, are not affordable for rural areas with weak economic conditions (Zheng et al., 2020), making these facilities prone to failure in operation at a later stage due to the lack of funds.

iii Select a process with simple and less operation and maintenance. There are various rural sewage stations widely distributed in rural areas, lacking sewage treatment professionals for operation and maintenance. Therefore, facilities that are simple in operation and maintenance without frequent process testing or adjustment are essential in rural sewage treatment (Wang, 2021a, 2021b).

iv. Select a process that is conducive to resource utilization. Sewage resource utilization means the sewage meets specific water quality standards upon harmless treatment and can be used as reclaimed water for residents' living, ecological water supply, and agricultural irrigation in replacement of conventional water resources, or other resources and energy can be extracted from sewage. It is of great significance to increase the supply of water resources, alleviate the shortage of irrigation water, and ensure the safety of water ecology.

v. Adjustment should be made as per local conditions. Field investigation should be conducted before the selection of the rural sewage treatment process. Based on the functional requirements of the receiving water, the treatment process suitable for the local area can be determined with consideration of the economic conditions of rural areas, the complete status of infrastructure, natural environment, rural water-using habits, water consumption, the permanent population, climatic conditions as well as sewage situation and the final drainage destination of surrounding factories and aquaculture farms.

2.4.2 Thought of Technical Selection for Rural Sewage Treatment

Based on the above technology selection principle, and the actual needs and conditions of rural domestic sewage treatment, the following technical processes should be adopted (Administration and for Market Regulation, 2019):

i. With the purpose of removing organic matters, sewage can be discharged or recovered as resources upon being treated by bio-contact oxidation units.

ii. With the purpose of removing organic matters, sewage can be discharged or recovered as resources upon passing by the anaerobic organism membrane unit and biological treatment unit in the areas qualified for setting biological units.

iii. With the purpose of removing TN, sewage can be discharged or recovered as resource after being treated by anoxic and oxic biological units.

iv. With the purpose of removing TN and TP, sewage can be discharged or recovered as resources by phosphorus removal units after being treated by anoxic and oxic biological units.

2.4.3 Integrated Sewage Treatment

Based on the main principles and selection thought for the rural sewage treatment process discussed above, sewage treatment plants must be set and constructed to meet the requirements of low investment, high quality, and convenient operation and maintenance due to the shortage of funds for rural infrastructure construction and the low technical level and management level of employed person in specific practices. More than that, the setting of independent and closed process units should be considered since the outbreak of the COVID-19 pandemic in 2020, reducing the contact between sewage and management personnel. The integrated sewage treatment process becomes mainstream in the implementation of the current process based on the above characteristics.

Based on biochemical reactions, the integrated sewage treatment plant is a sewage treatment assembly formed in the factory through the organic combination of various functional units such as pretreatment, biochemical, sedimentation, disinfection, and sludge return with electrical components, instrument components, pipelines, automatic control systems, and equipment rooms (Wen, 2016). Biofilm processes such as biological contact oxidation, and biological filter as well as activated sludge processes such as A/O, and A^2/O, or a combination of two or more processes can be adopted as the core biochemical unit. The integrated sewage treatment plant with standardized production conforms to the principles of complete sets, modularization, and automation. Its process, structural and appearance design, test processes, and type inspection are regulated by relevant standards to ensure the equipment quality.

The integrated process and structural design can save civil construction costs with a compact facility structure and small floor area. Also, standard production can ensure reliable operation, and simple operation and maintenance, so that operation and maintenance personnel can carry out the daily operation after simple training. Moreover, as the integrated facility is assembled in the factory and transported as a whole, with simple on-site civil construction, it can be automatically operated after installation and commissioning, boasting the advantages of short construction time and low construction cost. Operation and maintenance are performed more conveniently using intelligent control and online monitoring technology, effectively saving labor costs. These features make the integrated sewage treatment plant particularly suitable for rural domestic sewage treatment, and it has been extensively promoted in rural sewage treatment, and grown into a mainstream treatment process.

A Japanese Johkasou is a typical integrated sewage treatment plant. The combined process of anaerobic filter-contact oxidation is adopted as the mainstream process of the combined treatment Johkasou in Japan (Ministry of the Environment Protection, 2011). China introduced Johkasous from Japan by the end of the 20th century and has carried out a series of localized technology and equipment research and development based on the characteristics of rural domestic sewage (Wang, 2021a). The integrated treatment processes and facilities of varying scales have been developed for different single households, connected households, villages, and towns, which have been extensively applied and achieved satisfactory results.

Moreover, the integrated sewage treatment process and facilities have been continuously innovated and developed from numerous applications and practices. Meanwhile, progress has been made in the improvement of the main process, the optimized combination of the process flow, and the improvement of the carrier performance in recent years. The integrated sewage treatment plant will play a crucial role in the fields of rural sewage treatment and decentralized sewage treatment in the future.

References

Ai, C., & Cui, D. (2018). Research progress in treatment of village and town wastewater by oxidation ditch process. *Liaoning Chemical Industry, 47*(6), 533–535.

Cao, R., Wang, B., & Gao, G. (2004). Study on the mechanism of organic matter removal at ECO-pond of Dongying. *Journal of Hebei Institute of Architectural Science and Technology, 21*(3), 14–17.

Chan, R. T., & Stenstrom, M. K. (1979). *Use of the rotating biological contactor for appropriate technology wastewater treatment.* Water Resources Program, School of Engineering and Applied Science.

Cheng, S. K., Li, Z. F., Uddin, S. M. N., et al. (2018). Toilet revolution in China. *Journal of Environmental Management, 216*, 347–356.

China Association for Engineering Construction Standardization. (2019). *Code for design of building water supply and drainage: GB 50015-2019.* China Architecture & Building Press.

China Urban Water Association. (2021). *Outline of 2035 urban water industry development plan.* China Architecture & Building Press.

Comeau, Y., Hall, K. J., Hancock, R. E., et al. (1986). Biochemical model for enhanced biological phosphorus removal. *Water Research, 20*(12), 1511–1521.

Cui, C., Pan, K., & Shi, Z. (2021). Study on anaerobic biofilm reactor for advanced treatment of rural domestic sewage. *Technology of Water Treatment, 47*(6), 104–109.

Dai, X. (2020). Development analysis of municipal sludge treatment and disposal industry in China. *Science, 72*(6), 4, 30–34.

Deng, Y. H., & Wheatley, A. (2016). Wastewater treatment in Chinese rural areas. *Asian Journal of Water, Environment and Pollution, 13*(4), 1–11.

Escudié, R., Cresson, R., Delgenès, J. P., et al. (2011). Control of start-up and operation of anaerobic biofilm reactors: An overview of 15 years of research. *Water Research, 45*(1), 1–10.

Fan, B., Wang, H., Zhang, Y., et al. (2017). Application and development of septic tank technology in decentralized wastewater treatment. *Chinese Journal of Environmental Engineering, 11*(3), 1314–1321.

Gao, F., & Li, M. (2018). Application of rotating biological contactors in rural sewage treatment. *China Resources Comprehensive Utilization, 36*(1), 56–58.

Han, X. (2011). *A simulated experimental research on treating rural domestic sewage through a stabilization pond process.* Northeast Agricultural University.

Hang, S., Liu, X., & Liang, P. (2004). Misunderstanding of sludge disposal and treatment and control strategy. *China Water and Wastewater, 20*(12), 89–92.

Hao, X., Chen, Q., Li, J., et al. (2018). Status and trend of MBR process application in the world. *China Water and Wastewater, 34*(20), 7–12.

Henze, M., Van, L. M. C., Ekama, G. A., et al. (2008). *Biological wastewater treatment.* IWA.

Hu, Q. (2011). *Study on treatment of biological contact oxidization and greenhouse structure subsurface-flow constructed wetland for processing rural domestic sewage.* Harbin Institute of Technology.

Islam, M. S. (2017). Comparative evaluation of vacuum sewer and gravity sewer systems. *International Journal of System Assurance Engineering and Management, 8*(1), 37–53.

Jia, S. (2016). *Research on biologically pretreated coal gasification wastewater treatment by two stage membrane bioreactor.* Harbin Institute of Technology.

Kadlec, R. H., & Wallace, S. (2008). *Treatment wetlands* (2nd ed.). CRC Press.

Kamel, M. M., & Hgazy, B. E. (2006). A septic tank system: On site disposal. *Journal of Applied Sciences, 6*(10), 2269–2274.

Karadag, D., Köroğlu, O. E., Ozkaya, B., et al. (2015). A review on anaerobic biofilm reactors for the treatment of dairy industry wastewater. *Process Biochemistry, 50*(2), 262–271.

Krzeminski, P., van der Graaf, J. H., van Lier, J. B., et al. (2012). Specific energy consumption of membrane bioreactor (MBR) for sewage treatment. *Water Science and Technology, 65*(2), 380–392.

Li, P., Sun, Y., Sui, K., et al. (2021). Analysis on the present situation and discussion on its treatment mode of rural sewage treatment in China. *Water and Wastewater Engineering, 57*(12), 65–71.

Li, X., Ding, A., Zheng, L., et al. (2018). Application of constructed wetlands for water pollution treatment in China during 1990–2015. *Environmental Engineering, 36*(4), 5, 11–17.

Lin, S. (2015). *A comprehensive evaluation for municipal sewage plant based MBR technology.* Tsinghua University.

Liu, H. (2019). *Environmental protection equipment—Principle, design and application.* Chemical Industry Press.

Liu, Y., Kumar, S., Kwag, J. H., et al. (2013). Magnesium ammonium phosphate formation, recovery and its application as valuable resources: A review. *Journal of Chemical Technology and Biotechnology, 88*(2), 181–189.

Mahapatra, S., Samal, K., & Dash, R. R. (2022). Waste stabilization pond (WSP) for wastewater treatment: A review on factors, modelling and cost analysis. *Journal of Environmental Management, 308*, 114668.

Matsuo, T. (2015). *Water environment engineering.* Ohmsha.

Meng, F. G., Chae, S. R., Drews, A., et al. (2009). Recent advances in membrane bioreactors (MBRs): Membrane fouling and membrane material. *Water Research, 43*(6), 1489–1512.

Ministry of Environmental Protection. (2011). *Technical specifications of the biological contact oxidation sewage treatment works: HJ 2009–2011.* China Environmental Press.

Ministry of Housing and Urban-Rural Development of the People's Republic of China, State Administration for Market Regulation. (2019). *Technical standard for domestic wastewater treatment engineering of rural area: GB/T 51347—2019.* China Architecture & Building Press.

Mino, T., Loosdrecht, M., & Heijnen, J. J. (1998). Microbiology and biochemistry of the enhanced biological phosphate removal process. *Water Research, 32*(11), 3193–3207.

Mino, T., & Matsuo, T. (1984). Principal mechanism of biological phosphate removal. *Water Pollution Research, 7*, 605–609.

Ogawa, H., & Iwahori, K. (2002). Removal characteristics of suspended solids and organic compounds in anaerobic filter process of small-scale domestic wastewater treatment facilities (Gappei-shori Johkasou). *Journal Japan Biological Society of Water and Waste, 38*(2), 69–77.

Patsios, S. I., & Karabelas, A. J. (2010). A review of modeling bioprocesses in membrane bioreactors (MBR) with emphasis on membrane fouling predictions. *Desalination and Water Treatment, 21*(1–3), 189–201.

Ren, J. (2005). *Application of chlorine dioxide disinfection in sewage treatment plant.* Tianjin University.

Satou, Y. (2013). Structure, function and maintenance management of advanced treatment septic tank incorporating iron electrolytic phosphorus removal equipment. *Johkasou, 448*, 10–16.

Shiohara, T., Ebie, Y., Kakiki, A., et al. (2005). Inactivation effect of UV-led on indicator bacteria in Johkasou effluent water. *Proceedings of JSCE G (Environment), 76*(7), III_243–III_250.

Sopper, W. E., & Kardos, L. T. (1973). *Recycling tread municipal wastewater and sludge through forest and cropland.* Pennsylvania State University Press.

Sun, J. Y., Liang, P., Yan, X. X., et al. (2016). Reducing aeration energy consumption in a large-scale membrane bioreactor: Process simulation and engineering application. *Water Research, 93*, 205–213.

Tchobanoglous, G., & Schroeder. (1985). *Water quality: Characteristics, modeling, modification.* Addison Wesley Pub. Co.

Thakre, S. B., Bhuyar, L. B., & Deshmukh, S. J. (2009). Oxidation ditch process using curved blade rotor as aerator. *International Journal of Environmental Science & Technology, 6*(1), 113–122.

Wang, B. (2019). *Purification performance of odorous-black water and it's microbial mechanisms in integrated constructed wetland.* Harbin Institute of Technology.

Wang, H. (2018). Exploring the way of rural sewage treatment in China—Planning, construction and management of rural sewage treatment facilities. *Water and Wastewater Engineering, 54*(5), 1–3.

Wang, J. (2021a). *Effectiveness of Johkasou for treating rural dispersed domestic sewage.* Shanghai Normal University.

Wang, L. (2015). *Study on treatment for small town sewerage with the combination process of biological filter and biological contact oxidation tank.* Chongqing University.

Wang, T. X. (2021b). Study on rural domestic sewage treatment scheme. *IOP Conference Series: Earth and Environmental Science, 781*(3), 032040.

Wen, Y. (2016). *Systemic solutions for China's rural sewage treatment.* Chemical Industry Press.

Wu, H. (2014). *Cyclic processes of carbon, nitrogen and phosphorus in constructed wetlands and its environmental effects.* Shandong University.

Wu, M., Zhang, D., Xu, S., et al. (2019). Research progress of dephosphorization technology on wastewater. *Nonferrous Metals Science and Engineering, 10*(2), 97–103.

Ye, M., Wang, P., Liu, Y., et al. (2021). Outdoor vacuum sewerage system and its application in rural sewage treatment engineering in China. *Journal of Environmental Engineering Technology, 11*(6), 1196–1201.

Yoshida, Y., Takahashi, K., Saito, T., et al. (2005). Nitrite inhibition of aerobic phosphate uptake alleviated by denitrifying activity of polyphosphate accumulating organisms. *Environmental Engineering Research, 42*, 69–79.

Yue, S., Liu, X., Shi, C., et al. (2013). Research and application progress of biofilter process for wastewater treatment and reuse. *Technology of Water Treatment, 39*(1), 1–6.

Zeng, J., Zhao, Y., Ji, M., et al. (2021). Risk control and disinfection measures of rural sewage treatment facilities under pandemic situations. *Journal of Water Resources and Water Engineering, 32*(3), 51–57.

Zhang, J., Xiao, K., & Huang, X. (2020). Full-scale MBR applications for leachate treatment in China: Practical, technical, and economic features. *Journal of Hazardous Materials, 389*, 122138.

Zhang, Q. (2021). *Research on rural domestic sewage treatment based on anaerobic biofilm coupled with ABR septic tank.* North China University of Water Resources and Electric Power.

Zhang, Y., Lyu, M., Xu, M., et al. (2021). Functional orientation of the three-chamber septic tank toilet and errors in its application in rural toilet reform. *Journal of Agricultural Resources and Environment, 38*(2), 215–222.

Zhang, Z. (2015). *Wastewater engineering.* China Architecture & Building Press.

Zheng, X., Gao, Y., Xu, Y., et al. (2022). Application status and model types of three-compartment septic tanks in rural toilet reform in China. *Journal of Agricultural Resources and Environment, 39*(2), 209–219.

Chapter 3
Integrated Treatment Technology for Domestic Sewage in Rural Areas

3.1 A^3/O-MBBR Process

3.1.1 R&D Background

A^2/O, as a classic sewage treatment process, has been extensively applied in rural sewage treatment. The A^2/O process is superior in efficiently removing organic matters, total nitrogen (TN) and total phosphorus (TP), but defective in its unstable effect of denitrification and phosphorus removal, which is not applicable for the treatment of sewage with low C/N and C/P. The short residence time of denitrification and long residence time of nitrification is the cause of unstable TN in water treated by the A^2/O process. Moreover, the nitrate nitrogen of reflux sludge affects the anaerobic environment, leading to a poor effect of biological anaerobic phosphorus release effect (Chen et al., 2019a, 2019b). Also, under the influence of living habits, C/N and C/P are low in most rural sewage, which may result in unsatisfactory nitrogen and phosphorus removing efficiency by the A^2/O process (Qin et al., 2018). Therefore, the A^3/O process is developed to further improve the efficiency of nitrogen and phosphorus removal by adding a pre-denitrification zone before the anaerobic zone. It returns the sludge to the pre-denitrification zone, thus removing nitrates while maintaining a perfect anaerobic environment in the back-end anaerobic zone and overcoming the defects related to the A^2/O process.

Moving Bed Biofilm Reactor (MBBR) is a derivative process of the biological contact oxidation process. It achieves efficient removal of pollutants through the carrier-attached biofilm by adding suspended carrier to the reactor (Yang et al., 2017). The carrier is suspended in water as its density approaches to that of water, providing a place for developing the attachment of microorganisms, as shown in Fig. 3.1. It is conducive to increasing the number and types of microorganisms in the system, offering a more favorable selection for functional microorganisms, and providing enrichment sites for long-generation microorganisms such as nitrifying bacteria. The carrier has a re-cutting effect on air bubbles, which can improve the utilization

© The Author(s) 2024
W. Li et al., *Integrated Treatment Technology of Rural Domestic Sewage*,
https://doi.org/10.1007/978-981-99-5906-8_3

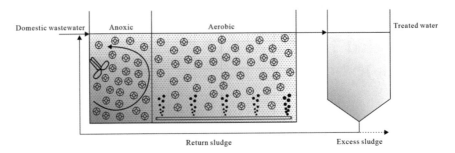

Fig. 3.1 MBBR process flow (Zhou et al., 2021)

of oxygen in the system. The fluidized state of the carrier should be achieved via aeration or stirring in most cases to intensively mix the carrier with sewage.

A^3/O-MBBR integrated treatment process is the "pre-denitrification-anaerobic-anoxia- aerobic (MBBR)-precipitation-soft carrier filtration" combination process. It is characterized by the improvement in the anaerobic environment at the back end through adding the pre-denitrification zone in front of the anaerobic zone, and the enhanced biomass and species of the system as well as the shortened residence time of the aerobic zone through adding the suspended carriers to the aerobic zone. Further, SS can be efficiently removed through combing partial precipitation and soft fixed carrier filtration. And A^3/O-MBBR can also be flexibly designed as a mud-film composite process or a full-scale biofilm process as per different water inlet conditions.

3.1.2   Process Flow

The flow of the A^3/O-MBBR (mud-film composite) integrated treatment process is shown in Fig. 3.2. Domestic sewage collected by the pipeline flows through the pre-denitrification zone, the anaerobic zone, the anaerobic tank and the aerobic zone successively for biochemical treatment. Microorganisms undergo a denitrification reaction with nitrate nitrogen in the return sludge in the pre-denitrification zone using organic matters in raw water. Hydrolysis of organic matters and biological phosphorus release is achieved in the aerobic zone. Nitrogen removal by denitrification can be achieved by denitrifying bacteria in the anaerobic tank where the dissolved oxygen concentration is extremely low. Meanwhile, partial alkalinity provided by denitrification can create favorable conditions for the subsequent nitrification reaction in the aerobic zone. The preferred biomass-increasing sponge carrier in the aerobic zone has the advantages of high sludge concentration (with MLSS reaching up to 10,000–20,000 mg/L), high volume load, strong resistance to impact load, low sludge yield and long service life. Microorganisms in activated sludge and carrier can decompose organic matter under aerobic conditions. Also, organic nitrogen and ammonia nitrogen are gradually converted into nitrite and nitrate via the nitrification

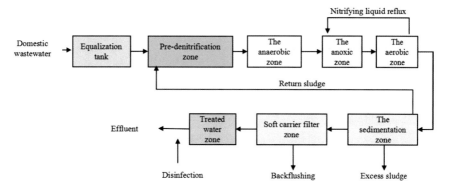

Fig. 3.2 Process flow of A³/O-MBBR (mud-film composite)

reaction, and nitrification liquid is refluxed from the aerobic zone to the anaerobic tank. At the same time, high-concentration phosphorus-containing sludge is produced with excessive phosphorus uptake in phosphorus accumulating bacteria. Finally, mud and water are separated in the sedimentation zone, and the supernatant is filtered and sterilized by soft fixed carriers to meet the standard for discharging or recycling. The excess sludge in the sludge hopper of the secondary secondary sedimentation tank is discharged into the sludge thickening tank for concentrating and drying before transportation for treatment.

The A³/O-MBBR (full-scale biofilm) integrated treatment process has main functional areas including the anoxic zones 1, 2, and 3, the aerobic zone, secondary sedimentation tanks and soft carrier filtration, as shown in Fig. 3.3. High-efficiency fixed carriers are set in the anoxic zone, with the preferred biomass-increasing sponge carrier added to the aerobic zone.

The full-scale biofilm process is adopted for treating domestic sewage with low concentration without adding suspended active sludge. When treating conventional or high-concentration sewage, activated sludge in a suspended state can be added to improve the efficiency. The full-scale biofilm process is superior in terms of strong resistance to impact load, few excess sludges as well as simple operation and maintenance, but should be complemented with other means to address its defects in low phosphorous removing efficiency.

3.1.3 Design Parameters

Design water volume: 30–500 m³/d.

Design parameters of the A³/O-MBBR (mud-film composite) process: The residence time of the biochemical section: 10.4 h; The gas–water ratio: 17:1; the reflux ratio of sludge: 100%; the reflux ratio of nitrifying liquid: 200%; MLSS: 2000–3500 mg/L; the dissolved oxygen in the aerobic zone: 2–4 mg/L.

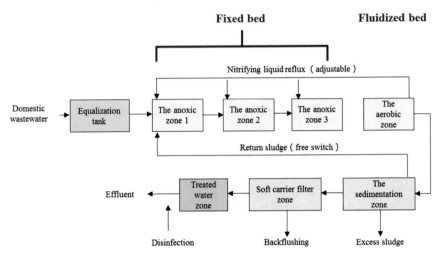

Fig. 3.3 Process flow of A^3/O-MBBR (full-scale biofilm)

Design parameters of the A^3/O-MBBR (full-scale biofilm) process.

The residence time of the biochemical section: 12.6 h; the gas–water ratio: 50:1–70:1; the reflux ratio of sludge: 100%; the reflux ratio of nitrifying liquid: 200–400%; MLSS: less than 1000 mg/L; the dissolved oxygen of the multi-stage aerobic zone: greater than 5 mg/L.

Design parameters for the quality of domestic sewage and treated water are shown in Table 3.1.

Table 3.1 Design parameters for the quality of domestic sewage and treated water

	COD$_{Cr}$ (mg/L)	BOD$_5$ (mg/L)	NH$_3$–N (mg/L)	TN (mg/L)	TP (mg/L)	SS (mg/L)	pH
Quality of domestic sewage (mud-film composite)	400	200	40	50	5	200	6–9
Quality of domestic sewage (full-scale biofilm)	200	100	25	30	3	200	6–9
Quality of treated water	50	10	5 (8)	15	0.5	10	6–9

Note The value outside the brackets is the control index when the water temperature is greater than 12 °C while the value inside is the control index when the water temperature is smaller than or equal to 12 °C

3.1.4 Features

a. Prominent effect of removing nitrogen and phosphorus with stable effluent.
b. Minimized carbon source is added in the anoxic zone, with low operating cost.
c. The residence time in the aerobic zone can be significantly reduced, saving land occupation and reducing energy consumption.
d. Filtration with soft fixed carriers can lower energy consumption without the need for filter pumps and backwash pumps.
e. The risk of sludge bulking can be effectively avoided.
f. A complete food chain is formed in the system, effectively reducing sludge volume.
g. The carrier has strong resistance to impact load and a long service life.

3.1.5 Scope of Application

A^3/O-MBBR (mud-film composite) process can be extensively applied in rural sewage treatment, domestic sewage treatment in scenic spots without municipal pipe networks, decentralized domestic sewage treatment in schools, hospitals, and inns as well as the source control and sewage interception for black and odorous water. It is especially applicable for areas with strict nitrogen and phosphorus discharge standards, and sewage treatment with low C/N and C/P. The A^3/O-MBBR (full-scale biofilm) process is mainly applicable for low-concentration domestic sewage treatment.

3.1.6 Tips for Operation and Maintenance

a. TN and TP removal rates can be enhanced by 5–10% by distributing part (70–95%) of the domestic sewage into the anaerobic zone in the A^3/O-MBBR (mud-film composite) process.
b. When operating A^3/O-MBBR (mud-film composite) process in winter, the MLSS in the system can be appropriately increased.
c. Short flow in the anoxic zone should be avoided in the A^3/O-MBBR (full-scale biofilm) process.
d. Intermittent aeration should be established in the anoxic area/anaerobic zone to prevent sludge settling and excessive biofilm thickness, with regular inspection of its normal operation.
e. The carrier in the aerobic zone should be checked for no accumulation, no damage, and no non-functional biological overload.
f. The biomass-increasing sponge carrier selected requires no replacement within 5 years. After that, 5% of the carriers should be supplemented every two years.

g. The carrier block at the end of the aerobic zone should be unobstructed, and cleaned once every 15–30 days if necessary.

h. The clean water tank should be cleaned at unregulated intervals, normally 30–60 days.

3.2 Improved Bardenpho-MBBR Process

3.2.1 R&D Background

The Bardenpho process is a typical process for efficiently conducting simultaneous denitrification, as shown in Fig. 3.4. The process is formed by adding an anoxic zone and an aerobic zone to the A/O process, so it is also known as a four-zone enhanced denitrification process. Efficient denitrification is the main strength of the Bardenpho process. With abundant carbon sources, domestic sewage in the first anaerobic tank has a high denitrification efficiency. Denitrifying bacteria mainly denitrify with intracellular carbon sources through endogenous respiration to improve the denitrification effect in the second anaerobic tank. It is defective in poor phosphorus removing effect. Hence, an anaerobic zone is added in front of the first anaerobic tank to improve its phosphorus removal efficiency, forming an improved Bardenpho process, as shown in Fig. 3.5.

The improved Bardenpho-MBBR integrated treatment process is a combined process of "anaerobism-anoxia-MBBR-anaerobism-anoxia-sediment-filtration with soft carriers". This process, combining advantages of the modified Bardenpho process and the MBBR process, is characterized by strong impact resistance, high TN loads, outstanding comprehensive treatment performance, and favorable treatment water quality (meeting the environmental quality standard of surface water).

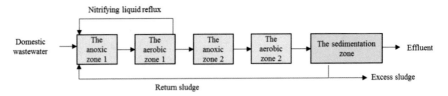

Fig. 3.4 Flow of Bardenpho process (Wang, 2019)

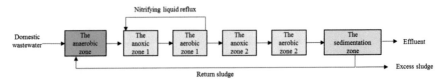

Fig. 3.5 Process flow of improved Bardenpho-MBBR integrated treatment

3.2.2 Process Flow

The improved Bardenpho-MBBR integrated treatment process is shown in Fig. 3.5. Domestic sewage collected by the pipeline flows through the anaerobic zone, the anoxic zone 1, the aerobic zone 1, the anoxic zone 2, and the aerobic zone 2 for biochemical treatment. On this basis, mud and water are separated in the sedimentation zone, and the supernatant is filtered by soft fixed carriers and sterilized to meet the standard for discharging or recycling. The excess sludge in the sludge hopper of the secondary secondary sedimentation tank is discharged into the sludge thickening tank for concentrating and drying before transportation for treatment.

Main mechanisms of different biochemical treatment zones:

a. The anaerobic zone is functioned as decomposing organic matter and releasing phosphorus with the help of phosphorus accumulating bacteria. The efficiency of phosphorus release with phosphorus accumulating bacteria is greatly improved in the strictly anaerobic environment. Also, the enhanced biodegradability of sewage is conducive to the subsequent aerobic treatment.
b. High-efficiency denitrification of nitrate nitrogen in the backflow nitrification solution can be achieved in the anoxic zone 1 by microorganisms with the help of organic matters in water. Meanwhile, partial alkalinity offered by denitrification can provide a favorable condition for the subsequent nitrification in the aerobic zone.
c. The preferred biomass-increasing sponge carrier in the aerobic zone 1 has the advantages of high MLSS (up to 10,000–20,000 mg/L), high volume load, strong resistance to impact load, low sludge yield, and long service life. The microorganisms in activated sludge and carrier decompose organic matter under aerobic conditions. Also, organic nitrogen and ammonia nitrogen are gradually converted into nitrite and nitrate via the nitrification reaction, and nitrification liquid is refluxed from the aerobic zone to the anaerobic tank. At the same time, phosphorus accumulating bacteria complete excessive phosphorus uptake, producing high-concentration phosphorus-containing sludge.
d. The anoxic zone 2 uses the externally added carbon source or the carbon source provided by the diverted inlet to conduct a denitrification reaction on nitrate nitrogen contained in the treated water of the anoxic zone 1, thus improving the denitrification efficiency of the system.
e. The aerobic zone 2 removes the residual organic matter, ammonia nitrogen, and fine air bubbles attached to the sludge flocs in the treated water of the anoxic zone 2.

3.2.3 Design Parameters

Design water volume: 50–500 m^3/d.

Design parameters: The residence time of the biochemical section: 12.6 h.

Table 3.2 Design parameters for the quality of domestic sewage and treated water

Water quality indexes	COD_{Cr} (mg/L)	BOD_5 (mg/L)	NH_3–N (mg/L)	TN (mg/L)	TP (mg/L)	SS (mg/L)	pH
Domestic sewage	400	200	45	50	5	250	6–9
Treated water	30	6	1.5 (2.5)	15	0.3	10	6–9

Note The value outside the brackets is the control index when the water temperature is greater than 12 °C while the value inside is the control index when the water temperature is smaller than or equal to 12 °C

The gas-water ratio: 25:1–30:1; the reflux ratio of nitrifying liquid: 150–200%; the reflux ratio of sludge: 50–100%.

MLSS: 2500–5000 mg/L.

Design parameters for the quality of domestic sewage and treated water are shown in Table 3.2.

3.2.4 Features

a. There is a prominent effect of comprehensive water treatment with high TN loads, with stable and high-quality effluents.
b. Minimized carbon source is added in the anoxic zone according to the step feed, reducing operating cost.
c. The slashed residence time in the aerobic zone can save land occupation and reduce energy consumption.
d. Filtration with soft fixed carriers can lower energy consumption without the need for filter pumps and backwash pumps.
e. The risk of sludge bulking can be effectively avoided.
f. A complete food chain is formed in the system, effectively reducing sludge volume.
g. The carrier has strong resistance to impact load, and a long service life.

3.2.5 Scope of Application

The improved Bardenpho-MBBR integrated process can be applied in rural sewage treatment, domestic sewage treatment in scenic spots without municipal pipe networks, decentralized domestic sewage treatment in schools, hospitals, and inns as well as the source control and sewage interception for black and odorous water. It is especially applicable for sewage treatment with a high nitrogen load (TN ≤ 70 mg/L).

3.2.6 Tips for Operation and Maintenance

a. Part of the raw water should be distributed into the anoxic zone 2, which can lower the amount of external carbon sources, saving operation and maintenance costs.
b. The MLSS in the system should be appropriately enhanced during operation in winter.
c. Intermittent aeration can be set in the anaerobic zone/anoxic zone to prevent sludge sedimentation; also, regular inspection should be performed to ensure normal operation.
d. The carrier in the aerobic zone should be checked for no accumulation, no damage, and no non-functional biological overload.
e. The carrier block at the end of the aerobic zone 1 should be unobstructed, which can be cleaned once every 15–30 days if necessary.
f. The biomass-increasing sponge carrier selected requires no replacement within 5 years. After that, 5% of the carriers should be supplemented every two years.
g. When the carbon source is added to the anoxic zone 2, the sludge growth of the system will be accelerated with the rising demand for daily sludge discharged.
h. The clean water tank should be cleaned at unregulated intervals, normally 30–60 days.

3.3 Multi-level and Multi-stage A/O-MBBR Process

3.3.1 R&D Background

The denitrification efficiencies of traditional A/O and A^2/O processes are limited by the reflux ratio of nitrification solution, and their theoretical removal rate of TN can only reach up to 70%, making them not suitable for rural domestic sewage treatment with a high TN concentration. The step-feed multi-level multi-stage A/O process (referred to as SMMAO process) is a derivative process of the single-level and single-stage A/O activated sludge process. Two segments of A/O processes are added to the typical A/O process (Liu et al., 2012), with the theoretical removal rate of TN improved up to 78% (on the premise of being free from reflux of nitrification solution). The As SMMAO process can effectively improve the TN removal rate of the system, therefore it is appropriate for the treatment of rural domestic sewage with high TN concentration. But it still has a large room for improvement in the nitrification rate, tank capacity, and SS concentration of treated water. SMMAO-MBBR integrated process is the combined process of "anoxic zone 1-aerobic zone 1 (MBBR)-anoxic zone 2-aerobic zone 2 (MBBR)-anoxic zone 3-aerobic zone 3 (MBBR)-precipitation-soft carrier filtration". Its feature is that the preferred biomass-increasing sponge carrier is added to each aerobic zone on the basis of the SMMAO process, and based on the advantages of the MBBR process, it can strengthen the nitrification and denitrification efficiency and improve the comprehensive water treatment efficiency of the system.

3.3.2 Process Flow

The integrated process flow of SMMAO-MBBR is shown in Fig. 3.6. Domestic sewage collected by the pipeline flows through the anoxic zone 1, the aerobic zone 1, the anoxic zone 2, the aerobic zone 2, the anoxic zone 3, and the aerobic zone 3 in segments. On this basis, mud and water are separated in the sedimentation zone, and the supernatant is filtered by soft fixed carriers and sterilized to meet the standard for discharging or recycling. The excess sludge in the sludge hopper of the secondary secondary sedimentation tank is discharged into the sludge thickening tank for concentrating and drying before transportation for treatment.

Main mechanisms of different biochemical treatment zones:

a. High-efficiency denitrification of nitrate nitrogen in the backflow nitrification solution can be achieved by the microorganisms in the anoxic zone 1 with the help of organic matters in water. At the same time, partial alkalinity offered by denitrification can provide a favorable condition for the subsequent nitrification in the aerobic zone.

b. The nitrate nitrogen generated in the treated water of the aerobic zone 1 is denitrified in the anoxic zone 2 using the carbon source provided by the step feed domestic sewage.

c. The nitrate nitrogen generated in the treated water of the aerobic zone 2 is denitrified in the anoxic zone 3 using the carbon source provided by the step feed domestic sewage.

d. The microorganisms in the preferred biomass-increasing sponge carrier and the activated sludge can decompose organic matters under aerobic condition in the aerobic zones 1, 2, and 3. Organic nitrogen and ammonia nitrogen are converted into nitrite and nitrate via nitrification reaction.

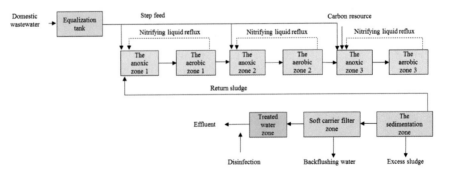

Fig. 3.6 Step-feed multi-level and multi-stage A/O-MBBR process flow

Table 3.3 Design parameters of sewage inlet and treated water under SMMAO-MBBR process

Water quality indexes	COD$_{Cr}$ (mg/L)	BOD$_5$ (mg/L)	NH$_3$–N (mg/L)	TN (mg/L)	TP (mg/L)	SS (mg/L)	pH
Quality of domestic sewage	500	300	80	100	5	200	6–9
Quality of treated water	50	20	5 ($\leq$8)	15	0.5	10	6–9

Note The value outside the brackets is the control index when the water temperature is greater than 12 °C while the value inside is the control index when the water temperature is smaller than or equal to 12 °C

3.3.3 Design Parameters

Design water volume: 50–500 m^3/d.

Design parameters: The residence time of the biochemical section: 20.4 h; The gas-water ratio: 20:1–30:1; the reflux ratio of nitrifying liquid: 50–100%; the reflux ratio of sludge: 50–100%; MLSS: 2500–5000 mg/L.

Design parameters for the quality of domestic sewage and treated water are shown in Table 3.3.

3.3.4 Features

a. A high TN removing efficiency, and stable and high-quality effluents.
b. The system has a small tank volume, reducing roughly 25% of the occupied area compared with the multi-level single-stage A/O system.
c. Minimized carbon source is added in the anoxic zone by step feeding with equal inlet distribution.
d. The slashed residence time in the aerobic zone can save land occupation and reduce energy consumption.
e. The TN removal rate can be improved by up to 90% with the reflux of nitrifying liquid in each A/O zone.
f. Filtration with soft fixed carriers can lower energy consumption without the need for filter pumps and backwash pumps.
g. The risk of sludge bulking can be effectively avoided.
h. A complete food chain is formed in the system, effectively reducing sludge volume.
i. The carrier has strong resistance to impact load and a long service life.

3.3.5   Scope of Application

The SMMAO-MBBR process can be used for the treatment of rural domestic sewage with high TN concentration (TN $\leq$ 100 mg/L) in domestic sewage as well as decentralized sewage treatment in areas such as expressway service areas, some public toilets, and scenic spots.

3.3.6   Tips for Operation and Maintenance

a. The reflux ratio (0–100%) of nitrifying liquid should be flexibly adjusted according to the domestic sewage, saving operation energy consumption and maintenance costs.
b. When C/N is insufficient, carbon sources should be added to control C/N within 4:1 and 6:1.
c. The MLSS in the system can be appropriately enhanced during operation in winter.
d. Intermittent aeration should be established in the anoxic zone to prevent sludge settling, with regular inspection of its normal operation.
e. The carrier in the aerobic zone should be checked for no accumulation, no damage, and no non-functional biological overload.
f. The carrier block at the end of the aerobic zone should be unobstructed, which can be cleaned once every 15–30 days if necessary.
g. The biomass-increasing sponge carrier selected requires no replacement within 5 years. After that, 5% of the carriers should be supplemented every two years.
h. The clean water tank should be cleaned at unregulated intervals, normally 30–60 days.

3.4   Multi-stage A/O Biological Contact Oxidation Process

3.4.1   R&D Background

The A/O activated sludge process has become one of the mainstream sewage treatment processes because of its simple process, simultaneous denitrification and removal of carbon, and high treatment efficiency. But it is defective in poor resistance to water quality and impact load, and high excess sludge production (Du et al., 2017), with the lack of professional operation and maintenance management in rural domestic sewage treatment. To deal with the non-acclimatization of the A/O process in rural domestic sewage treatment, a multi-stage A/O process (multi-stage A/O biological contact oxidation process) was developed based on the biofilm method.

The process is to carry out subdivision on the A and O segments of the traditional A/O process forming multi-stage anoxic zones (A1, A2, A3) and aerobic zones (O1, O2, O3), which also adds anoxic and aerobic carriers to the anoxic zone and the aerobic zone, respectively, based on the theory of biotic population in ecology (Xue, 2020). To be specific, the multi-stage anoxic zone adopts the form of a fixed bed, while the multi-stage aerobic zone adopts a fixed bed or the fluidized bed (MBBR) process. Based on the process, the dominant bacteria in each zone are more prominent and the system biomass increases through the biofilm on the carrier, enhancing the system's resistance to water quality and water impact.

3.4.2 Process Flow

The process flow of multi-stage A/O biological contact oxidation integrated treatment is shown in Fig. 3.7. Domestic sewage collected flows into the pretreatment unit (solid–liquid separation tank), the multi-stage anoxic zone, the multi-stage aerobic zone, and the soft carrier filtration zone successively before meeting the standard for discharging.

Main mechanisms of various biochemical treatment zones:

a. The solid-liquid separation zone has the comprehensive effect of regulating water volume, solid-liquid separation, and reduced dissolved oxygen (deoxidizing).
b. The multi-stage anoxic zone contributes to SS interception, hydrolysis of organic matters, and denitrification, with varied effects in each stage. The anoxic zone 1 functions as interception, precipitation, and deoxidation, and the anoxic zones 2 and 3 are responsible for the hydrolysis and denitrification of organic matters.
c. The multi-stage aerobic zone functions as the degradation and nitrification of organic matter, among which the aerobic zone 1 degrades organic matters, and the aerobic zones 2 and 3 realize the nitrification of ammonia nitrogen and organic nitrogen.

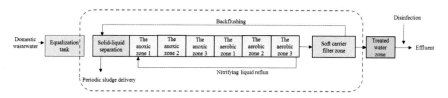

Fig. 3.7 Process flow of multi-stage A/O biological contact oxidation integrated treatment

Table 3.4 Design parameters of sewage inlet and treated water under multi-stage A/O integrated treatment

Water quality indexes	COD_{Cr} (mg/L)	BOD_5 (mg/L)	NH_3–N (mg/L)	TN (mg/L)	TP (mg/L)	SS (mg/L)	pH
Quality of domestic sewage	500	300	80	100	5	200	6–9
Quality of treated water	50	20	5 (≤8)	15	0.5	10	6–9

Note The value outside the brackets is the control index when the water temperature is greater than 12 °C while the value inside is the control index when the water temperature is smaller than or equal to 12 °C

3.4.3 Design Parameters

Design water volume: 0.6–50 m^3/d.

Design parameters: The residence time of the biochemical section: 26–32 h.

The gas-water ratio: 60:1–70:1; the reflux ratio of the mixture: 200–400%; the filling ratio of the multi-stage anoxic zone: 50–70%; the filling ratio of the multi-stage aerobic zone: 30%; the dissolved oxygen of the multi-stage aerobic zone: 6–8 mg/L.

Design parameters for the quality of domestic sewage and treated water are shown in Table 3.4.

3.4.4 Features

a. The system uses biofilm to efficiently remove pollutants and has a strong resistance to impact load.

b. A solid-liquid separation zone may not be set as per the process requirements.

c. Anoxic zone 1 can effectively eliminate oxygen and provide anoxic conditions beneficial to denitrification for the anoxic zones 2 and 3.

d. The carbon source addition is minimized with improved utilization of organic matters in the multi-stage anoxic zone, saving the operating cost.

e. The low concentration of organic matter in the aerobic zones 2 and 3 are conducive to the growth of nitrifying bacteria on the biofilm, improving nitrification rate.

f. Low sludge production (2‰), low sludge treatment and disposal cost, and long operation and maintenance period.

g. The system is simple with low requirements for operation and maintenance technology, saving the operation energy consumption and maintenance costs.

h. A precipitation zone should be set when the fixed bed process is adopted.

i. When the fluidized bed process is adopted, the MLSS can be reduced to lower than 100 mg/L, and no separate secondary secondary sedimentation tank is required in the system, reducing land occupation and simplifying operation and maintenance.

3.4.5   Scope of Application

The multi-stage A/O biological contact oxidation integrated treatment process is extensively applied in rural domestic sewage treatment, especially for domestic sewage and tail water treatment in scenes such as villages, inns, farmhouses, villas, and scenic spots.

3.4.6   Tips for Operation and Maintenance

a. This process should be supplemented with phosphorus removal measures to achieve effective phosphorus removal.
b. The appropriate water temperature is 10–25 °C; insulation can be performed through burying or adding a thermal insulation layer if necessary.
c. The carrier in the multi-stage anoxic zone requires no replacement and supplement normally.
d. There requires no replacement of carriers when the fixed bed process is adopted in the multi-stage aerobic zone. And updating and maintenance can be considered after five years of adopting the fluidized bed process.
e. The anoxic zone should be cleaned and dug periodically. In general, the top layer floats should be cleaned first, and then the sludge at the bottom is pumped.
f. The carrier block in the multi-stage aerobic zone should be unobstructed for selecting the fluidized bed process or should be cleaned if necessary.
g. When the carrier in the multi-stage anoxic zone is blocked, it can be dredged using aeration.
h. Carbon sources can be supplemented to make the TN compliance rate higher when the C/N ratio of domestic sewage is low.
i. The dissolved oxygen in the reflux liquid can be appropriately lowered by adjusting the aeration parameters to enhance the denitrification efficiency.

Fig. 3.8 SND-type
biological contact oxidation
model

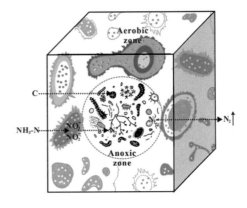

3.5   SND-Type Biological Contact Oxidation

3.5.1   R&D Background

Simultaneous Nitrification and Denitrification (SND) refers to the biological denitrification process that is synchronized in time and space of nitrification and denitrification reactions under the conditions of no significant anoxic and aerobic partitions in space and no anoxic/aerobic alternation in time (Yang et al., 2003). The process has outstanding advantages such as a simple process, short residence time, and small floor space, but it is defective in its limited denitrification efficiency (Run et al., 2012).

The SND-type biological contact oxidation process is a process to achieve simultaneous nitrification and denitrification based on special carriers of certain volume and biofilms grown on the carriers. Its specific implementation is shown in Fig. 3.8. A certain number of preferred porous sponge carriers placed in the SND zone may generate a DO gradient inside and outside of the carrier due to the limitation of oxygen diffusion. The outer surface of the carrier is dominated by aerobic nitrifying bacteria with high dissolved oxygen, and deep inside the carrier, denitrifying bacteria are dominant since an oxygen-deficient zone is generated due to the obstruction of oxygen transfer. In that case, the microenvironment generated in the carrier contributes to achieving simultaneous nitrification and denitrification.

3.5.2   Process Flow

The process flow of SND-type biological contact oxidation is shown in Fig. 3.9. The domestic sewage collected flows into the system and reaches the SND zone after the pretreatment of the solid–liquid separation zone 1 and zone 2 (with the effect of regulating water volume), which then passes through the SND1-SND3 zones in

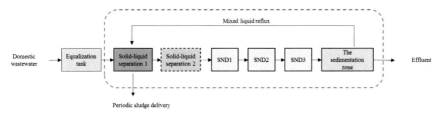

Fig. 3.9 Process flow of SND-type biological contact oxidation

turn. Then, efficient biomass increasing sponge carrier for SND is placed in the tank to realize the simultaneous degradation of organic matters, ammonia nitrogen and TN under the synergistic action of various microorganisms on the carrier surface and inside the carrier. Mud-water separation can be achieved upon treated water in the SND3 zone flows into the sedimentation zone. Of which, the lower inorganic mud is air-lifted and refluxed to the solid–liquid separation zone according to the set reflux ratio, and the supernatant is discharged or recycled after being sterilized.

3.5.3 Design Parameters

Design water volume: 0.6–50 m³/d.

Design parameters: The residence time of the biochemical section: 14.4–16 h; the gas-water ratio: 50:1–60:1; the reflux ratio of the mixture: 200–400%; The filling ratio of the biomass incremental sponge carrier specific for SND: 30%; the dissolved oxygen in the SND zone: 3–6 mg/L.

Design parameters for the quality of domestic sewage and treated water are shown in Table 3.5.

Table 3.5 Design parameters for the quality of domestic sewage and treated water

Water quality indexes	COD$_{Cr}$ (mg/L)	BOD$_5$ (mg/L)	NH$_3$–N (mg/L)	TN (mg/L)	TP (mg/L)	SS (mg/L)	pH
Quality of domestic sewage	400	200	40	50	5	200	6–9
Quality of treated water	60	20	5 (8)	30	1	20	6–9

Note The value outside the brackets is the control index when the water temperature is greater than 12 °C while the value inside is the control index when the water temperature is smaller than or equal to 12 °C

3.5.4 Features

a. SND process can be achieved, avoiding the inhibition of NO_3^- accumulation on the nitrification reaction, accelerating the rate of the nitrification reaction.
b. The alkalinity released by the denitrification reaction can partially compensate for the alkali consumption of the nitrification reaction, making the pH value of the system stable.
c. The process is simple without the reflux of nitrifying liquid, reducing cost and difficulty in operation and maintenance.
d. The floor space is small, the tank capacity can be reduced by 20–30%.
e. The denitrification efficiency of SND is approximately 50%, lower than that of the traditional A/O process.
f. The fluidized bed or fixed bed SND process can be flexibly selected.
g. With a small amount of mud produced, it has a long maintenance period.

3.5.5 Scope of Application

The SND-type biological contact oxidation process is applicable for household/combined domestic sewage treatment and small centralized domestic sewage treatment in rural areas with low limits ($\leq$30 mg/L) or no limits for TN discharge.

3.5.6 Tips for Operation and Maintenance

a. This process should be supplemented with phosphorus removal measures to achieve effective phosphorus removal.
b. The appropriate water temperature is 10–25 °C; insulation can be performed through burying or adding a thermal insulation layer if necessary.
c. The selected biomass increasing sponge carrier for SND requires no replacement within eight years. Maintenance is required after eight years.
d. The inorganic sludge in zone 1 of solid-liquid separation should be cleaned on a regular basis (ranging from 3 to 6 months).
e. The carrier block in the SND zone should be unobstructed for selecting the fluidized bed process, which should be cleaned once every 15–30 days if necessary.

3.6 A/O-MBR Process

3.6.1 R&D Background

The A/O activated sludge process is advantageous in the efficient removal of organic matters and TN. To ensure a sufficient nitrification reaction, a longer residence time should be set in the aerobic zone, with a risk of poor treated water quality caused by sludge bulking. The organic combination of the process and membrane bioreactor (MBR) is widely used in the field of sewage treatment (Huang et al., 2020), which can intercept microorganisms and other suspended solids (incl. bacteria, viruses, and insect eggs.) through setting a hollow fiber membrane or flat membrane (membrane pore size 0.01–1 μm) in the biochemical reaction tank and replace the secondary secondary sedimentation tank in the activated sludge process to achieve solid–liquid separation (Hong, 2022).

The A/O-MBR integrated treatment process is the combined process of "pretreatment zone—anoxic zone—aerobic zone (MBR)—treated water zone". It is characterized by the short process flow, short residence time, no risk of sludge bulking, good quality of treated water, and direct recycling of treated water (Li, 2007, 2016).

3.6.2 Process Flow

The A/O-MBR integrated processing process flow is shown in Fig. 3.10. The sewage is intercepted by a fine grid machine for removing hair and coarse particles to achieve the pretreatment effect before flowing through the anoxic zone and the aerobic zone for biochemical treatment. After that, it enters the treated water zone for discharging or recycling. Denitrification and degradation of some organic matters are performed in the anoxic zone. High MLSS in the aerobic zone can effectively degrade organic matters and carry out nitrification reactions, and the mixed liquid is returned to the anoxic zone; the membrane module can filter and intercept refractory organic matters and sludge. The process can regulate the volume ratio of the anoxic zone to the aerobic zone, therefore effectively cutting the biochemical residence time, reducing the equipment size, and lowering investment costs.

3.6.3 Design Parameters

Design water volume: 30–500 m^3/d.

Design parameters: The residence time of the biochemical section: 8 h.

The gas-water ratio: 50:1–60:1; the reflux ratio of the mixture: 200–400%; MLSS: 8000–12,000 mg/L.

Dissolved oxygen in the aerobic zone: 2–4 mg/L.

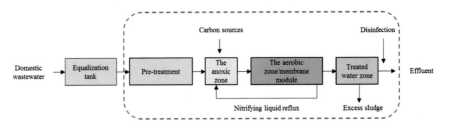

Fig. 3.10 A/O-MBR process flow

Table 3.6 Design parameters for the quality of domestic sewage and treated water

Water quality indexes	COD_{Cr} (mg/L)	BOD_5 (mg/L)	NH_3-N (mg/L)	TN (mg/L)	TP (mg/L)	SS (mg/L)	pH
Quality of domestic sewage	400	200	40	50	5	350	6–9
Quality of treated water	20	10	5 (8)	15	0.5	10	6–9

Note For the index outside the brackets of ammonia nitrogen, the water temperature is greater than or equal to 12 °C; for the index in the brackets of ammonia nitrogen, the water temperature is less than 12 °C

Of which, the size of the gas-water ratio is associated with the membrane material, type, production design and the process flow. The membrane air-sweeping water ratio can reach 3:1–3.5:1 with the advances of technology. And the biochemical tank can be separated from the membrane tank, with the gas-water ratio of the process reaching from 10:1 to 12:1, or even smaller.

Design parameters for the quality of domestic sewage and treated water are shown in Table 3.6.

3.6.4 Features

a. There is a prominent effect of SS treatment with stable and high-quality water production, and recycling.
b. The process is simple with the short flow, short residence time and small occupation of land.
c. With high MLSS, the system has a strong resistance to impact load.
d. With a small amount of mud produced, the treatment cost is low.
e. No risk in sludge bulking.
f. Membrane modules are expensive with high operation and maintenance costs.

g. The process is complex with high technical requirements for operation and maintenance, and is normally combined with intelligent control.
h. The process can be flexibly designed as per the working conditions, and the aerobic zone can be separated from the membrane zone.

3.6.5 Scope of Application

The A/O-MBR process is applicable for rural domestic sewage treatment in areas with developed economies, land constraints, and water resources shortages as well as treatment of decentralized domestic sewage from scenic spots, schools, hospitals, and inns in environmentally sensitive areas without municipal pipeline networks.

3.6.6 Tips for Operation and Maintenance

a. The appropriate water temperature is 10–25 °C; insulation can be performed through burying or adding a thermal insulation layer if necessary.
b. When C/N is insufficient, carbon sources should be added to control C/N within 6:1 and 8:1.
c. The MLSS in the system can be appropriately enhanced during operation in winter.
d. Intermittent aeration should be established in the anoxic zone to prevent sludge settling, with regular inspection of its normal operation.
e. It shall regularly check the pressure gauge, flow meter and turbidity meter of the effluent system, as well as any damage to the membrane module.
f. Membrane cleaning is the highlight of the operation and maintenance of the process, including daily backwashing with clean water and regular online cleaning of chemicals. Daily backwashing with treated water (for example, backwash once for 1–2 min every 6 h or 24 h) is performed according to the quality of raw water and the situation of produced water. When the membrane pressure difference is high (the maximum limit of the hollow fibrous membrane and the flat membrane is 60 kPa and 20 kPa, respectively), the cleaning agent such as NaClO with the concentration of 1–3‰ should be used for online cleaning (the membrane biological process water treatment engineering technical specification for sewage treatment HJ 2010–2011). In general, the hollow fiber membranes should be cleaned no less than once a month, and the flat membranes can be cleaned once every 2–3 months.

3.7 Membrane-Aerated Biofilm Reactor

3.7.1 R&D Background

The biofilm process is a common sewage treatment process featuring strong adaptability to changes in water quality and quantity, and convenient management. It removes pollutants in sewage by the dense biofilm attached to the carrier surface. COD, NH_3–N and other pollutants in sewage can enter the biofilm through diffusion, and are then decomposed and removed by a variety of aerobic bacteria, anaerobic bacteria and other microorganisms in the biofilm (Henze et al., 2008). Normally, oxygen, COD, and NH_3–N are diffused and transferred from the outside to the inside of biofilm in the same direction (Semmens et al., 2003). Therefore, autotrophic nitrifying bacteria are normally disadvantageous in competition with heterotrophic aerobic bacteria on the biofilm surface, resulting in limited treatment efficiency of NH_3–N and TN. On the other hand, the biological treatment process, in general, supplies oxygen using aeration with large power consumption and large cost accounting for 60–80% of the total operating cost (Sun, 2015). At the same time, the short residence time of bubble generated by traditional aeration in water results in low oxygen utilization efficiency, less than 20% (Ahmed & Semmens, 1992). Membrane Aerated Biofilm Reactor (MABR) technology has been highly concerned for its superiority over the above two aspects. MABR is a bubble-free aeration technology that treats sewage using a breathable membrane and attached biofilm. In general, hollow fiber microporous membranes or dense silicone rubber membranes with hydrophobic properties are adopted as the breathable membrane in MABR (Sun, 2015). During aeration, air enters the water body in the form of dissolved diffusion or extremely tiny bubbles, thus a high oxygen utilization rate can be obtained with the aeration power efficiency reaching 10 kgO_2/kWh (Sun, 2015). In the meantime, oxygen and pollutants diffuse in the biofilm in opposite directions, which are different from the forms of diffusing oxygen and pollutants in traditional biofilms, as shown in Fig. 3.11. Nitrifying bacteria have an advantage in the aerobic layer since ammonia nitrogen with small molecules is easier to diffuse compared with organic matter. On this basis, the removal of NH_3–N and TN in sewage can be enhanced by the simultaneous nitrification and denitrification (SND) together with the denitrifying bacteria in the anoxic layer. MABR technology is less applied in municipal sewage treatment worldwide with few cases for rural sewage treatment (Chen et al., 2019a, 2019b).

3.7.2 Process Flow

The process flow is shown in Fig. 3.12. Sewage enters the anaerobic-MABR reaction zone for biochemical treatment via pretreatment units such as fine grids, and then undergoes the separation of mud and water in the sedimentation zone. The supernatant

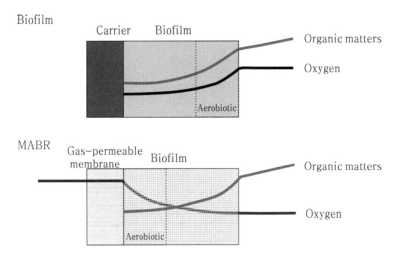

Fig. 3.11 Diffusion patterns of oxygen and organic matter in traditional biofilms and MABR biofilms

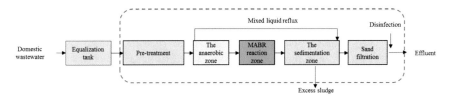

Fig. 3.12 MABR process flow

in the sedimentation zone can remove SS via the sand filtration system, and then it is sterilized by sodium hypochlorite before discharging. With the concrete structure as its main body, the system features an integrated layout and a compact layout as a whole.

3.7.3 Design Parameters

Design water volume: 20–100 m³/d.

Design parameters: The residence time in the MABR reaction zone: 0.38–18 h; the air flow rate: 1.4–5.3 L/m² h (Lu et al., 2021), and the mixture reflux ratio: 50–100%.

Design parameters for the quality of domestic sewage and treated water are shown in Table 3.7.

Table 3.7 Design parameters for the quality of domestic sewage and treated water

Water quality indexes	COD_{Cr} (mg/L)	BOD_5 (mg/L)	NH_3-N (mg/L)	TN (mg/L)	SS (mg/L)	pH
Quality of domestic sewage	400	200	40	50	200	6–9
Quality of treated water	50	10	5 (8)	15	10	6–9

Note The value outside the brackets is the control index when the water temperature is greater than 12 °C while the value inside is the control index when the water temperature is smaller than or equal to 12 °C

3.7.4 Features

Bubble-free aeration, heterogeneous mass transfer and layered structure are the main features of MABR (Wei, 2012). MABR supplies oxygen in the form of membrane aeration. Oxygen enters the water body through the polymer membrane in the bubble-free form and is then utilized by microorganisms attached to the surface of the aeration membrane. Theoretically, the oxygen utilization can reach 100%. Moreover, oxygen and pollutants diffuse into the biofilm in opposite directions, which is beneficial to the formation of the aerobic layer, anoxic layer, and anaerobic layer in the biofilm as well as the simultaneous nitrification and denitrification reactions, realizing a high denitrification efficiency. MABR process is characterized by:

a. With the prominent oxygen utilization, roughly 50–80% of the aeration volume can be saved, lowering energy consumption.
b. Outstanding BOD and TN treatment effects.
c. The system can achieve simultaneous nitrification and denitrification.
d. The alkalinity released by the denitrification reaction can partially compensate for the alkali consumption of the nitrification reaction, making the pH value of the system stable.
e. The floor space is small, the tank capacity can be reduced by 20–30%.
f. The process is simple, which can turn off the reflux of nitrifying liquid according to the conditions of domestic sewage and treated water, reducing costs and difficulty in operation and maintenance.
g. With a small amount of mud produced, the treatment cost is low.
h. The membrane module is expensive, with high construction costs as well as operation and maintenance costs.
i. Low Noise and no Odor.

3.7.5 Scope of Application

Membrane modules of MABR are expensive. In general, the price of a set of membrane boxes is as high as several hundred thousand yuan. Hence, the high cost will be the main factor limiting its large-scale application. For this reason, MABR is suitable for rural domestic sewage treatment in areas with a developed economy, high land occupation requirements, and high TN concentration of domestic sewage.

3.7.6 Tips for Operation and Maintenance

The removal efficiency of ammonia nitrogen shows a downward trend with the rising biofilm thickness, according to relevant research findings (JIEI Innovation Laboratory). Hence, reasonably controlling biofilm thickness is a focus of operation and maintenance as per the concentration of domestic sewage and the standard of treated water.

a. The membrane module should be flushed periodically through aeration to reasonably control the biofilm thickness.
b. The normal function of the pretreatment unit should be checked periodically, to prevent hair from entering the system and damaging the membrane components.
c. Check any damage to the breathable membrane regularly.
d. The appropriate water temperature is 10–25 °C; insulation can be performed by burying or adding a thermal insulation layer if necessary.
e. This process should be supplemented with phosphorus removal measures to achieve effective phosphorus removal.
f. The inorganic sludge in the sedimentation zone should be cleaned regularly (ranging from 3 to 6 months).

3.8 Improved Anaerobic Biofilm Process

3.8.1 R&D Background

Water resources on earth are approximately 1.4 billion km^3, of which roughly 97.5% are seawater, with only 0.01% of the freshwater existing in the form of rivers and lakes (easy utilization), showing the scarcity of freshwater resources on earth (Ministry of Land, Infrastructure, Transport and Tourism, 2021). Studies suggest that two-thirds of the world's population may face water scarcity by 2025, and roughly half of the world's population may suffer from high water stress by 2030 with global population growth, expansion of industrial and agricultural activities, and global warming (Scheierling et al., 2011).

Reclaimed water is an unconventional water resource (resource utilization of sewage), its utilization is one of the effective solutions to alleviate water shortage (Almuktar et al., 2018). Given that about 70% of the world's water is used for irrigation (Pedrero et al., 2010), farmland irrigation becomes an important direction for sewage resource utilization. In China, for instance, the total agricultural water consumption was 368.23 billion m^3 in 2019, accounting for roughly 61.2% of the national water consumption. 100 billion m^3 of sewage produced nationwide can greatly ease the pressure on farmland irrigation if they are recycled (Ministry of Water Resources, 2019).

Rural domestic sewage is relatively simple in composition because of no mixture of industrial sewage, which can meet the requirements of farmland irrigation upon simple biological treatment. For the decentralized collection and decentralized treatment of rural domestic sewage, as shown in Table 2.1, septic tanks are normally adopted to simply treat black water for resource utilization. But the amount of water flowing into septic tanks has risen sharply with the popularization of flushing toilets, resulting in the residence time of feces being lower than the design value and poor quality of effluent. To enhance the treatment capacity of septic tanks, up-flow anaerobic sludge bed reactor (UASB) septic tanks and carrier-based septic tanks have been developed (Fan et al., 2017). The UASB septic tank with an upward flow inlet can enhance the removal rate of SS and dissolved organic matters, but it faces the problem of poor resistance to impact load. The carrier septic tank, in general, is filled with carriers such as ceramsite, elastic three-dimensional carrier, and gravels, the impact load resistance and removal of organic matters are improved in accordance with the anaerobic biofilm principle, but there is a problem of easy blockage.

The improved anaerobic biofilm integrated treatment process combines the strengths of the UASB septic tank and carrier septic tank, can enhance the removal efficiency of pollutants by setting the upward flow of the water inlet through the diversion tube. Meanwhile, the resistance to impact load of the system is strengthened using the anaerobic biofilm process after filling high-performance carriers. Further, the risk of carrier blocking can be lowered by setting the equipment at the back of the traditional septic tank.

3.8.2   Process Flow

The process flow of the improved anaerobic biofilm integrated treatment for sewage is shown in Fig. 3.13. Domestic sewage collected enters the septic tank to remove large solid particles and garbage, and then flows through the anaerobic zones 1 and 2 in the form of gravity flow for biochemical treatment. Finally, the treated water is utilized as a resource locally. The upward flow of domestic sewage is adopted in the anaerobic zone 1. To be specific, the sewage enters the lower part of the anaerobic zone through the guide tube and then flows through the preferred carrier layer during upward flow. Also, SS in sewage is intercepted efficiently with the help of the physical interception of carrier and the adsorption of biofilm. Meanwhile, organic matter is decomposed by

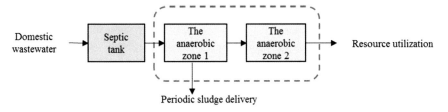

Fig. 3.13 Process flow of improved anaerobic biofilm

Table 3.8 Design parameters for domestic sewage and treated water of improved anaerobic biofilm process

Water quality indexes	COD_{Cr} (mg/L)	BOD_5 (mg/L)	SS (mg/L)	pH
Quality of domestic sewage	400	200	220	6–9
Quality of treated water	150	60	80	6–9

the anaerobic biofilm. In addition, no carrier is filled at the bottom and upper part of the zone for it is reserved for storing the deposited SS, the fallen biofilm, and the scum accumulated in the upper part. The anaerobic zone 2 adopts downward inflow water, that is, domestic sewage flows through the carrier layer from top to bottom, through which, organic matters are decomposed via anaerobic biofilm. Among them, carriers in the anaerobic zones 1 and 2 are of different sizes and shapes, which simultaneously possess large specific surface area, high bioburden, large porosity, and high-durability plastic material carrier, contributing that the anaerobic zones 1 and 2 perform their respective functions.

3.8.3 Design Parameters

Design water volume: 0.2–5 m³/d.
   Design parameters: The residence time of the biochemical section: 2–5 days, the filling rate of anaerobic zone 1: 20–50%.
   Filling rate of anaerobic zone 2: 40–70%.
   Design parameters for the quality of domestic sewage and treated water are shown in Table 3.8.

3.8.4 Features

a. The treated water can be used as a resource locally.
b. The system has a strong resistance to impact load with the efficient removal of pollutants using the biofilm.

c. No energy consumption with low operating cost.
d. Simple operation and management as well as a long maintenance period.
e. Low sludge production, low sludge treatment and disposal cost as well as low operation and maintenance costs.
f. The equipment is free of secondary pollution such as noise.
g. Secondary secondary sedimentation tank is not required, occupying small floor space.
h. The low efficiency in the removal of nitrogen and phosphorus can retain the nitrogen and phosphorus nutrients in the sewage to the maximum extent.

3.8.5 Scope of Application

The water treated by the improved anaerobic biofilm process can meet water using standards for the paddy field and dryland crop in the "Water Quality Standard for Farmland Irrigation" (GB5084-2021). And it is applicable for the resource utilization of domestic sewage with decentralized collection and treatment in rural areas. The treated water can be used for vegetable gardens, small orchards, and small gardens on private plots or residential lands.

3.8.6 Tips for Operation and Maintenance

a. With a long acclimation period of anaerobic biofilm, it is normal to have high BOD in treated water within 150–200 days after the machine startup.
b. The sludge deposited at the bottom of the equipment and the scum accumulated on the top should be regularly discharged.
c. The inlet and outlet pipes of the anaerobic zones 1 and 2 are in the form of a tee to prevent a high concentration of methane and hydrogen sulfide in the equipment.
d. When the equipment is blocked and there is a large amount of floating mud on the upper, the blocked part should be checked with the blockage removed in time.
e. Smoking and open flames should be kept away from the equipment.

References

Ahmed, T., & Semmens, M. J. (1992). Use of sealed end hollow fibers for bubbleless membrane aeration: Experimental studies. *Journal of Membrane Science, 69*(1–2), 1–10.

Almuktar, S. A., Abed, S. N., & Scholz, M. (2018). Wetlands for wastewater treatment and subsequent recycling of treated effluent: A review. *Environmental Science and Pollution Research, 25*(24), 23595–23623.

Chen, J., Li, X., Li, C., et al. (2019a). Control of returned sludge and its impact on TP removal. *Environmental Protection Engineering, 37*(3), 218–219, 234.

Chen, W., Pan, C., Li, X., et al. (2019b). The application research on rural domestic wastewater by MABR. *Technology of Water Treatment, 45*(5), 126–128, 134.

Du, Z., Wang, Y., Zhang, K., et al. (2017). The optimal design and experiment study on multilevel anoxic oxic bio-film reactor. *Technology of Water Treatment, 43*(5), 96–99.

Fan, B., Wang, H., Zhang, Y., et al. (2017). Application and development of septic tank technology in decentralized wastewater treatment. *Chinese Journal of Environmental Engineering, 11*(3), 1314–1321.

Henze, M., van Loosdrecht, M. C. M., Ekama, G. A., et al. (2008). *Biological wastewater treatment: Principles, modeling and design.* IWA

Hong, W. (2022). Application of MBR process in rural domestic sewage treatment. *Energy Conservation and Environmental Protection, 3,* 91–92.

Huang, S. J., Pooi, C. K., Shi, X. Q., et al. (2020). Performance and process simulation of membrane bioreactor(MBR) treating petrochemical wastewater. *Science of the Total Environment, 747,* 141311.

Li, H. (2007). *Experimental study on low-strength domestic wastewater treatment by anoxic/aerobic membrane bioreactor.* Hohai University.

Li, S. (2016). *Application of MBR process in rural decentralized sewage treatment in Gui'an New Area.* Guizhou University.

Liu, S., Yang, X., Shi, F., et al. (2012). Analysis and research of anaerobic-oxic multilevel anoxic-oxic phosphorus and nitrogen removal technology. *Water and Wastewater Engineering, 48*(S1), 191–194.

Lu, D., Bai, H., Kong, F., et al. (2021). Recent advances in membrane aerated biofilm reactors. *Critical Reviews in Environmental Science and Technology, 51*(7), 649–703.

Ministry of Land, Infrastructure, Transport and Tourism. (2021). Present situation of water resources in Japan.

Ministry of Water Resources of the People's Republic of China. (2019). *China water resources bulletin 2019.* China Water & Power Press.

Pedrero, F., Kalavrouziotis, I., Alarcón, J. J., et al. (2010). Use of treated municipal wastewater in irrigated agriculture—Review of some practices in Spain and Greece. *Agricultural Water Management, 97*(9), 1233–1241.

Qin, H., Zhou, S., Zheng, Y., et al. (2018). Effect of ecological dialysis technology on phosphorus removal in domestic sewage treatment. *China Science and Technology Information, 5*(13), 85–86.

Run, Y., Xiao, Z., & Song, L. (2012). *New technology, process and equipment for water treatment.* Chemical Industry Press.

Scheierling, S. M., Bartone, C. R., Mara, D. D., et al. (2011). Towards an agenda for improving wastewater use in agriculture. *Water International, 36*(4), 420–440.

Semmens, M. J., Dahm, K., Shanahan, J., et al. (2003). COD and nitrogen removal by biofilms growing on gas permeable membranes. *Water Research, 37*(18), 4343–4350.

Sun, L. (2015). *The application of MABR technology in the ecological restoration of polluted urban river.* Tianjin University.

Wang, Y. (2019). Comparison between Bardenpho process and multilevel anoxic oxic process. *Tianjin Construction Science and Technology, 29*(6), 41–43.

Wei, X. (2012). *Research and application of new MABR technology for phosphorus and nitrogen removal.* Tianjin University.

Xue, C. (2020). *Research and application of A/O step-feed domestic sewage treatment process.* Lanzhou Jiaotong University.

Yang, Q., Li, X., Zeng, G., et al. (2003). Study progress on mechanism for simultaneous nitrification and denitrification. *Microbiology China, 4,* 88–91.

Yang, Y., Wu, D., Song, M., et al. (2017). Application of new MBBR in WWTP upgrading to meet Class IV surface water standard. *China Water and Wastewater, 33*(14), 93–98.

Zhou, Z., Gao, X., & Yang, D. (2021). Research on performance of MBBR technology for rural sewage treatment. *Yunnan Water Power, 37*(12), 205–207.

Chapter 4
Integrated Rural Domestic Sewage Treatment Equipment

Based on biochemical reactions, the integrated rural domestic sewage treatment plant is a sewage treatment assembly that is formed in the factory through the organic combination of various functional units such as pretreatment, biochemical, sedimentation, disinfection, and sludge return with electrical components, instrument components, pipelines, automatic control systems, and equipment rooms (Wen, 2016). The integrated equipment has been extensively promoted in practice as its outstanding advantages such as stable production quality, compact structure, small footprint, short construction period, low construction cost, reliable operation, and simple operation and maintenance, and it particularly conforms to the needs of rural domestic sewage treatment.

The integrated treatment equipment is divided into single household, joint household, village and town levels according to the treatment scale. The biofilm process and its derivative processes, featuring small investment, strong resistance to impact load, simple operation, and long operation and maintenance cycle, are selected for the decentralized household/joint-household integrated treatment equipment with the treated capacity less than 5 m^3/d, such as multi-stage A/O biological contact oxidation process, SND biological contact oxidation process, and anaerobic biofilm process. Village-level domestic sewage treatment can be also divided into small centralized treatment (5–50 m^3/d) and large centralized treatment (50–500 m^3/d). The combined process of activated sludge method, such as A^3/O-MBBR, improved Bardenpho-MBBR, and A/O MBR, is adopted in the large centralized integrated treatment equipment for its strong treatment capacity, small floor space, and stable operation. Small centralized integrated treatment equipment can flexibly select biofilm process or activated sludge process in accordance with the actual sewage treatment needs.

The integrated equipment is divided into ground type, buried type and semi-buried type according to different installation methods, of which the ground type and the underground type are dominant. Buried installation is popular because of its advantages over saving land, being less affected by temperature, no noise pollution, and no damage to the landscape. It is worth noting that the installation method should

© The Author(s) 2024
W. Li et al., *Integrated Treatment Technology of Rural Domestic Sewage*,
https://doi.org/10.1007/978-981-99-5906-8_4

be scientifically selected as per the local climate and the surrounding environment, rather than blindly pursuing the buried installation. In general, the ground or semi-buried type should be selected from the perspective of installation and maintenance, and the buried type should be selected for saving land. The integrated processing equipment has a significant difference in the appearance and structural design, main material, structural anticorrosion, and manufacturing process selection due to varied installation methods.

4.1 Ground Equipment

4.1.1 Appearance Design and Structural Design

The appearance of the integrated ground sewage treatment equipment is designed after the extensive market research, user requirement analysis, interpretation of domestic and foreign environmental protection policies, and design experiments by the industrial design team. Square boxes or cylindrical tanks are primarily adopted for their compactness in structure, small occupation of land, convenient transportation, and economical application. Moreover, the visual image is beautiful and generous with a strong sense of quality, and in harmony with the natural and living environment.

Structural stability and reasonable spatial layout should be highlighted in structural design. Structural stability under the operation of full water load should be considered prioritized to ensure that structural deformation is within a controllable range in terms of structural strength. Functional divisions should be clearly defined with clear pipelines regarding internal space layout, so as to avoid space waste and short-circuit flow. Meanwhile, the requirements of construction, operation, and maintenance should be also met.

4.1.2 Material of Main Part

Non-metallic materials and metal materials are the manufacturing materials of the integrated sewage treatment ground equipment: Non-metallic materials (or products) will suffer significant physical and chemical changes after long-term exposure to sunlight, wind and rain, high temperature, and severe cold, which may lead to a gradual decline in their performance, affecting its durability. Therefore, the ground equipment, in most cases, is made of metal materials, such as Q235 carbon structural steel, 304 stainless steel, Q355 and weathering steel. 304 stainless steel is less utilized for its cost. Its price is roughly three times as expensive as that of carbon steel. Given the mechanical properties, anti-corrosion properties, and price of comprehensive materials, weathering steel can be the main plate for ground equipment; Q355 can be used as the main component profile, and 304 stainless steel pipes and plates can be used in small quantities.

Weathering steel, or atmospheric corrosion-resistant steel, is a series of low-alloy steels between ordinary steel and stainless steel, which is composed of alloy components including corrosion-resistant elements such as copper and nickel. It can form an amorphous spinel oxide layer with a thickness of roughly 50–100 μm between the rust layer and the substrate. The dense oxide film firmly bonded to the base metal can prevent oxygen and water in the atmosphere from penetrating into the steel matrix, slowing down the corrosion development. In this way, the corrosion resistance of weathering steel can reach 2 to 8 times that of ordinary carbon steel. In general, it is used with a reinforced anti-corrosion coating with a smaller thickness of the sheet, reducing the weight and cost of the equipment.

4.1.3 Processing and Manufacturing

The manufacturing process of the integrated sewage treatment ground equipment is composed of material preparation, cutting, welding, drilling, surface treatment, assembly, testing inspection, and packaging. The main manufacturing techniques include casting, forging, bending, stamping, machining, cutting, mechanical assembly, welding, and surface treatment. Core processes of metal welding and surface treatment, as well as the mechanical assembly, testing, and inspection processes, are introduced as following.

a. Metal welding. Manual arc welding, submerged arc welding, gas shielded welding, and electro slag welding are commonly used welding techniques. The factory automation welding technique is dominated by the carbon dioxide gas shielded welding process. The welding principle of carbon dioxide gas shielded welding is that an arc is generated between the welding wire and the weldment during welding; the welding wire is automatically fed and melted by the arc to form molten droplets before entering the molten tank. Carbon dioxide gas is sprayed via the nozzle to surround the arc and molten tank, isolating air and protecting welding metals. This process features low welding cost, high production efficiency, simple operation, high crack resistance of the weld, and a wide application range, and is especially suitable for the welding of thin, medium, and thick plates of sewage treatment equipment.

b. Surface treatment. Pre-treatment and post-treatment are involved in equipment surface treatment. Specifically, pre-treatment mainly includes degreasing, rust removal (such as sandblasting), pickling, and phosphating (such as surface cleaning, removal of rust layer and oxide scale); and post-treatment is comprised of coating anti-corrosion, including painting or powder spraying. Coating is the most effective means of equipment anti-corrosion, which should be designed according to relevant standards. For example, the durability of anti-corrosion coatings should be investigated with the most severe atmospheric corrosion environment grade C5I and immersion environment Im3 according to the standard of GB/T30790 *Anti-corrosion Protection for Steel Structure Using Paint and Varnish Protective Coatings System*; and coating adhesion should be designed according to the standard of GB/T9286-1998 *Cross-Cross Test of Paint and Varnish Films*. The exterior of the equipment box can be coated with epoxy zinc-rich primer + modified epoxy paint + polyurethane top paint in the design of an anti-corrosion scheme with a mid-term anti-corrosion period of 10 years. And when the dry film thickness is 200 μm, it can effectively be isolated from moisture and oxygen, damaging the condition for corrosion. And the dry film thickness of 300 μm can be adopted inside the equipment with the modified epoxy paint with multi-layer polyamide curing.

c. Mechanical assembly. The assembly process is composed of partial assembly and final assembly. Partial assembly involves the assembly of critical components such as pipelines, electrical control systems, and functional equipment; and final assembly is the process of installing components and parts to form an integrated device. Assembly should be performed in strict accordance with the drawings and process regulations. Also, the key assembly procedures should be calibrated to reduce errors and deviations to ensure the performance and functions of the assembled products.

d. Test. A quality inspection should be performed on the integrated processing equipment before delivery according to the following points: (i) Appearance inspection is to confirm that no defect is found on the surface, including no blistering, no depression, no convex hull, and no obvious color difference with favorable flatness; (ii) Process inspection is to verify that no lack of glue, no water leakage, and no looseness can be observed in the pipeline and pipe fittings with normal valve switches, no failure, no damage, and correctly installed process equipment; (iii) A structural performance test is to confirm that the material thickness conforms to the design requirement; there is no leakage in the box or tank in the closed water test, and structural deformation meets the relevant standards; (iv) The electrical control system is tested to check that the electrical equipment is operated normally with the correct input of the control procedure.

4.2 Buried Equipment

4.2.1 Appearance and Structural Design

The integrated buried sewage treatment equipment is normally a vertical or horizontal circular tank since the mechanical property of the circular structure is superior to that of the square box structure, as shown in Fig. 4.1. There is no special requirement for visual aesthetics of the buried equipment.

Structural design requirements of the buried integrated treatment equipment are similar to those of the ground integrated treatment equipment as a whole, as shown

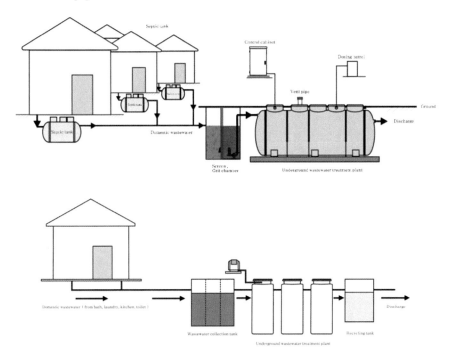

Fig. 4.1 Installation of integrated sewage treatment buried (vertical and horizontal) equipment

in 4.1.2, with a focus on structural stability and reasonable spatial layout. The over-burden pressure at no-load, the buoyancy generated by groundwater, the load of maintenance personnel, and the snow load should also be considered in the design of structural strength in addition to the structural stability under the operation of full water load. The appropriate ring stiffness is determined through deformation check calculation, strength calculation, and buckling instability calculation, to avoid excessively small ring stiffness of the equipment, excessive structural deformation and buckling instability caused by external pressure loads, or excessive ring stiffness and large section inertia moment, material waste and cost overspending.

4.2.2 Material of Main Part

The structure of the buried integrated treatment equipment is mainly made of non-metallic materials such as PPH, HDPE, and FRP (Table 4.1). Among them, FRP has been widely utilized in Japanese johkasous thanks to its high strength and low cost; PPH and HDPE, as renewable materials and environmentally friendly, have been widely used in recent years although they are expensive.

Table 4.1 Comparison of commonly used materials for buried equipment

Comparative project		PPH	PE	FRP
Density		Light	Light	Heavy
Corrosion resistant		Very good	Very good	Very good
Physical property	Pressure resistance	Good, suitable for burial installation	Good, suitable for burial installation	Very good, suitable for buried installation
	Heat insulation	Good	Good	Good
	Service life	More than 50 years	More than 50 years	More than 50 years
Appearance property		Exquisite appearance	Exquisite appearance	Poor appearance
Environment property		Renewable, non-toxic and odor-free, and comfortable manufacturing environment	Renewable, non-toxic and odor-free, and comfortable manufacturing environment	Non-renewable, toxic, strong odor, and poor manufacturing environment
Manufacturing process		High degree of automation in extrusion winding, injection molding and compression molding	High degree of automation in blow/injection/ rotational molding	Low degree of automation in winding, bonding and molding
Price		Expensive	General	Inexpensive

4.2.3   Processing and Manufacturing

The manufacturing process of the non-metal buried tank is composed of material preparation, cutting, molding, welding, drilling, surface treatment, mechanical assembly, testing inspection, and packaging. Of which, the processes of mechanical assembly, commissioning and inspection are similar to those in 4.1.3. The difference lies in the fact that welding and surface treatment is highlighted in the ground metal box manufacturing, while non-metal forming processes, such as extrusion winding, rotational molding, blow molding, winding bonding, and SMC molding are highlighted in non-metal tank manufacturing.

a. Extrusion winding. The process is extensively applied in the manufacturing of the PP material tank of large volume for the integrated equipment. It hot-melts the PPH pellets and molds them into a tank outside of the steel mold using a spiral extrusion winding unit. With a production efficiency 5–8 times higher than that of the manual process, its product has favorable performance, such as no joints, corrosion resistance, leakage resistance, and high aesthetics.

b. Rotational molding. It is a hollow molding method of thermoplastic, which is extensively used in the manufacturing of PE material tanks of large volumes for the integrated equipment. Plastic raw materials are firstly added to the mold that is continuously rotated and heated along two vertical axes. On this basis, the plastic raw materials are uniformly coated, melted, and adhered to the entire surface of the mold cavity under the action of gravity and thermal energy. After that, the product is obtained upon cooling and shaping. The process is characterized by low-cost tools and molds, flexible production, the capability of forming large and complex tanks, and a beautiful appearance.

c. Blow molding. It is a plastic treatment method that is developed rapidly and extensively applied in the manufacturing of PE material tanks of small volumes for the integrated equipment. The thermoplastic resin is extruded to form a tubular plastic parison during blow molding, which is placed in a split mold in a hot state (or heated to a softened state). Then, compressed air is injected into the parison immediately after closed molds, so that the parison is inflated and adhered to the inner wall of the mold. After that, various hollow tanks can be obtained upon cooling and demolding. The process is characterized by low-cost tools and molds, fast production, the capability of forming complex tanks, and a beautiful appearance.

d. Winding and bonding. It is one of the main manufacturing processes of resin-based composite materials. To be concrete, the fiber or cloth tape with resin glue is continuously, uniformly, and regularly winded on the core mold or lining with the special winding equipment under the condition of controlled tension and predetermined line shape, which is then finalized into a composite product of a certain shape in a certain temperature environment. It is widely used in the manufacturing of integrated equipment FRP material tanks of large volumes. And it is characterized by customized winding method, and customized winding law according to the stress condition of the product. The weight of winding vessels can be reduced by 40–60% compared with the steel vessels with the same volume and pressure. It also features low cost, and several materials (incl. resin, fiber and lining) can be selected and used in the same product to achieve the best economic effect.

e. SMC molding. This process is a process of firstly cutting the chopped fiber, resin, carrier and other sheets as per parameters such as product size, shape, thickness, and weight, then superimposing and placing them in the heated metal mold cavity, and finalizing and curing as per the set pressurizing method. It features simple operation, automation, a short production period (3–5 min), and the formation of products with smooth surfaces and complex structures. The manufactured products have excellent electrical insulation, mechanical properties, thermal stability, and chemical resistance.

Moreover, the structural performance test of the buried integrated treatment equipment should be carried out in a special sand pit in accordance with the *Small domestic sewage treatment equipment evaluation and certification rules* (T/CCPITCUDC-002-2021). The configuration of the anti-floating accessory must be considered in the equipment. Also, the inspection well must be locked and equipped with safety nets to prevent falling.

4.3   Automatic Control and Cloud Management System

The automatic control system is an automatic adjustment device or automatic program control device to control different process parameters, working states and production processes for a specific control object, which is in conformity with complex industrial control needs.

The "operation-free, low-maintenance" equipment requiring no special personnel to regularly adjust the process with the low difficulty in operation and maintenance is absolutely a need for rural sewage treatment plants with the characteristics of varied sites, discrete distribution, few professional operation and maintenance personnel, and low budget (Wang, 2018). In that case, the rural sewage integrated treatment equipment must be combined with digital means to ensure equipment operation, improve operation and maintenance efficiency and reduce costs.

Given the characteristics of the rural sewage integrated treatment equipment, the performance requirements of its automatic control system are halfway between civil and industrial applications. Also, simplified design should be performed on the premise of satisfying the requirements of process control, together with configurations such as standard data interfaces, communication protocol and communication network required for automatic control.

4.3.1   Traditional PLC Automatic Control System

a.   System Composition and Application Principle

PLC controller is the most commonly used logic control unit in the field of industrial automation control, and is widely used in the control system of municipal sewage treatment plants as well as parts of small and medium-sized sewage treatment devices.

Taking Hexu Chinese horizontal tank equipment as an example, the PLC system is constituted by a control cabinet and an external configuration. Specifically, the control cabinet is composed of key components such as circuit breakers, residual current protectors, lightning arresters, 24 V DC switching power supplies, PLC controllers, PLC analog expansion modules, switches, and gateways (Hexu or third-party), HMI touch screen, intermediate relays, thermal overload relays, current transformers, and fuses. The actuator with analog quantity and RS485 communication bus interface or sensors such as the liquid level gauge, the water quality sensor and the proportional valve are optional external configurations based on the process requirements. Standard Modbus interface protocols should be reserved in the control cabinet for third-party central control or cloud platform access control.

The principle of its equipment control system is shown in Fig. 4.2. With the PLC controller and switch as its core, parameter setting can be carried out via either the HMI touch screen or the third-party gateway or centralized control device. In the automatic control mode, the inlet liquid flow meter, the inlet temperature transmitter, and high-low liquid level float switches of the lifting tank work as input signals to

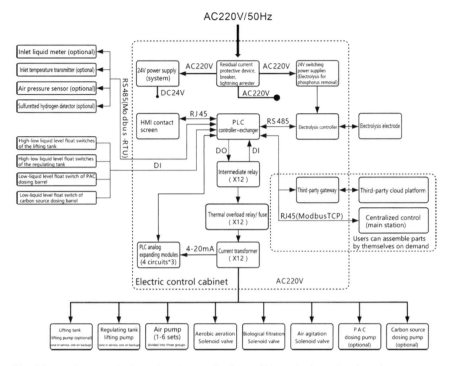

Fig. 4.2 Application principle of PLC controller in the Chinese horizontal tank equipment

output signals through the PLC controller to the lifting tank lifting pump and the equalization tank lifting pump for automatic operation. Residual current protectors, circuit breakers, and lightning arresters are utilized to ensure the electrical safety of the equipment.

Inlet liquid meter (optional).

Inlet temperature transmitter (optional).

Air pressure sensor (optional).

Sulfuretted hydrogen detector (optional).

24 V power supply (system); Residual current protective device, breaker, lightning arrester; 24 V switching power supplies (Electrolysis for phosphorus removal);

HMI contact screen; PLC controller + exchanger; Electrolysis controller; Electrolysis electrode;

High-low liquid level float switches of the lifting tank;

High-low liquid level float switches of the equalization tank;

Low-liquid level float switch of PAC dosing barrel;

Low-liquid level float switch of carbon source dosing barrel;

Intermediate relay;

Third-party gateway; Third-party cloud platform; Centralized control (main station);

Thermal overload relay/ fuse; current transformer; PLC analog expanding modules (4 circuits*3).

Lifting tank & lifting pump (optional); Equalization tank & lifting pump (one in service, one on backup).

Air pump (1–6 sets) divided into three groups.

Aerobic aeration solenoid valve; Biological filtration solenoid valve; Air agitation solenoid valve; PAC dosing pump (optional); Carbon source dosing pump (optional).

b. **Control Logic**

The control logic that should be realized in the control system of the integrated treatment equipment for rural domestic sewage includes:

Manual control, automatic control, semi-automatic commissioning, one in service and one standby control, group control, liquid level control, ultra-high liquid level protection, standby operation, interval operation, alternate operation between active and standby facilities, active-standby failover, on delay, off delay, and fault alarm.

4.3.2 *Internet of Things (IoT) Automatic Control System*

a. **R&D Background**

Although it is characterized by robust performance, simple programming, fast design, and ease to use (Zhang et al., 2017), the PLC controller has the following problems when applied to the rural sewage treatment equipment:

i. The PLC controller with a variety of discrete components should be customized for the wiring design and gateway configuration. Its high cost and large control cabinetsare unmatched by the small-tonnage sewage treatment equipment.

ii. The fault of the PLC electrical control cabinet should be handled by professional electrical engineers, rather than those operation and maintenance personnel with the basic knowledge of electricity.

iii. As there are normally two or more units acting as the equipment provider and the cloud platform provider, the owner shall closely connect with them to better understand the equipment principle, platform business, and the B&Rdging between equipment and platform (generally refers to the gateway, communication agreement involving two or more parties). Otherwise, the period, cost, quality, and performance of platform building cannot be guaranteed.

iv. Since there are many PLC manufacturers, large differences can be found in the existing PLC control system. In such a context, large costs will be spent for docking, rectification and commissioning during equipment informatization, which may lead to unsatisfactory monitoring effect.

v. At present, the small and medium-sized sewage treatment equipment on the market has a control complexity normally lower than that of conventional household appliances and industrial devices, but have high operation failure rates because of factors such as the complicated outdoor environment.

To address the above problems, it is of great necessity to develop an automatic control system for the small and medium-sized sewage treatment equipment in accordance with design requirements such as low cost, high space utilization, low failure rate, standardization, and informatization.

b. **Technical Scheme**

IoT automatic control technical scheme for varying application scenarios is developed for achieving automatic control with low cost, high space utilization, low failure rate, standardization, and informatization based on the Internet of Things (IoT) technology in combination with application scenarios, equipment technology, and operation and maintenance requirements as well as user's habit of decentralized sewage treatment equipment.

Concise system design is a must for the miniaturized application scenario. The Hexu H1 series IoT controller is taken as an example, with the main structure shown in Fig. 4.3. The system can meet the plug-and-play requirements of household/multi-household decentralized integrated sewage treatment equipment by integrating PLC, analog expansion modules, gateways, intermediate relays, current transformers, thermal overload relay, and electric energy meters, fuses, and even DC power supplies and electrical wiring.

Electric energy meter; Intermediate relay; current transformer; PLC analog expanding modules (4 circuits*3); Thermal overload relay (*3); Fuse; Electrical wiring;

H1 Series IoT Controller.

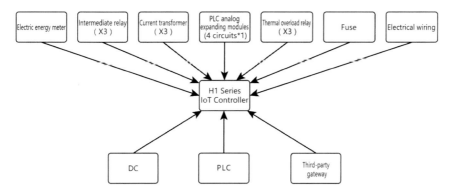

Fig. 4.3 Integration scheme of H1 series IoT controllers

DC; PLC; Third-party gateway.

The equipment control system that is simple, standard, easily maintained and expandable is essential for efficient operation and maintenance of sewage treatment equipment in large and medium-sized application scenarios such as village sewage treatment.

Hence, Hexu developed the X1 series "IoT gateway + output module + input module" combined technical solution that can be freely matched. Among them, the IoT gateway integrates PLC and third-party gateways; the output module integrates energy meters, intermediate relays, current transformers, PLC analog expansion modules, thermal overload relays, fuses and electrical wiring; and the input module integrates PLC, PLC digital input expansion module, and RS485 expansion, as shown in Fig. 4.4. The technical scheme can be flexibly adjusted according to the scale, characteristics, and requirements of the application scenario, which can effectively meet the automatic control requirements of small-scale/large-scale integrated sewage treatment equipment.

c. **Technical Features**

H1 series and X1 series IoT controllers are characterized by:

i. Robust performance. The controller integrates a 32-bit high-performance processor and rich communication interfaces and directly accesses the Internet through 4G/Cat1 and other mobile cellular networks to realize remote information monitoring. It also integrates power supply (H1 series), processor, 4G baseband, input, output, sensors, protectors, standardized control software, cloud platform access programs and other software and hardware, requiring no need for secondary development by users.

ii. Comprehensive functions. Manual control, automatic control, semi-automatic commissioning, one in service and one standby control, group control, liquid level control, ultra-high liquid level protection, standby operation, timing operation, interval operation, alternate operation between active and standby facilities, active-standby failover, on delay, off delay, voltage regulation, current

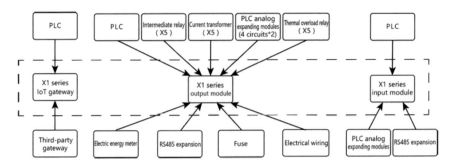

Fig. 4.4 Integration scheme of X1 series IoT controllers

regulation, voltage direction regulation, voltage measurement, current measurement, electric energy measuring, overload alarm, and open circuit alarm can be implemented in the equipment.

iii. Safe and reliable. The voltage, current and power of the control loop can be monitored in real-time with an intelligent diagnosis of abnormalities, which can prevent equipment failure or further damage to the load equipment, thus lowering the system failure rate.

iv. Universality. It can be extensively used in smart water affairs, smart agriculture, smart buildings, smart cities, and the fields that involve the control of IoT. The input voltage of the H1 series controller adopts the universe ultra-wide voltage range of 85–264 VAC considering the adaptability of the power supply.

d. **System Principle**

The electrical control system based on the X1 IoT controller of Hexu Chinese horizontal tank equipment is taken as an example. The system with AC220V/50Hz civil power supply can realize the joint control of lifting pump, air pump, solenoid valve, metering pump, liquid electromagnetic flowmeter, temperature sensor, gas pressure sensor, hydrogen sulfide detector and other devices or online monitoring instruments. It supports the full Netcom 4G/LTE Cat1 IoT communication network, and can directly access to the Hexu village sewage treatment management cloud platform by inserting the 4G data traffic card. Users can easily achieve remote monitoring of equipment via the platform's Web front-end or mobile Android end, and also conduct on-site monitoring and commissioning through the well-designed HMI touch screen interface. It can be operated without complex training. The principle of the whole control system is shown in Fig. 4.5.

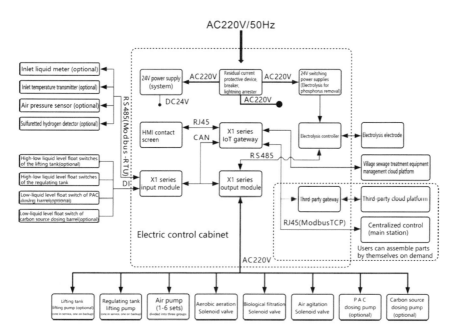

Fig. 4.5 X1 series Internet of Things controller in the Chinese horizontal tank equipment

X1 series IoT gateway, as the core control unit in the system, is responsible for logic control, data processing, and network communication. The X1 series input module can realize data acquisition of external switching devices and instruments, and the output module can achieve data acquisition for power control and current monitoring of external load equipment as well as the control of electrolytic dephosphorization drivers.

e. **Control Logic**

The electrical control system based on the X1 IoT controller in Hexu Chinese horizontal tank equipment is taken as an example. With simple process control requirements, the logic of basic function shown in Table 4.2 can be achieved through the IoT controller.

f. **Human–Machine Interface**

Human-machine interface (HMI) is the interface for the information interaction between humans and machines. The mainstream interface form is the touch screen interface that greatly simplifies the system complexity compared with the traditional physical buttons and regulating devices, improving efficiency of information interaction, and allowing users to better understand and control the device more intuitively. Normally, the HMI of the equipment automatic control system is composed of functions shown in Table 4.3.

The page design style of the HMI has no specific requirements, but should be conducted in accordance with the basic principles of simplicity and ease of use, flexible operation, clear logic, and complete functions. Examples of some functional blocks of the HMI are presented (Fig. 4.6).

g. **Data Interface**

There are a variety of manufacturers engaging in integrated treatment equipment for rural domestic sewage at home and abroad, with significant differences in their respective processes and technical schemes of automatic control systems.

The automatic control system should be designed with the industry-standard general communication interfaces, communication protocols, and data formats with an open data variable address table provided, so as to reduce data exchange barriers between different manufacturers' equipment as well as between equipment and third-party monitoring.

Industrial standard interfaces are adopted as communication interfaces, including RS232, RS485, RS422, CAN, and RJ45 (Ethernet).

Conventional industrial communication protocols should be adopted, of which MODBUS and TCP/IP protocols are the most commonly used, and the MQTT protocol is the most potential industrial IoT protocol in the current Industry 4.0 era.

The data format is a format for data exchange. Not much data is required in the integrated rural sewage treatment equipment. In principle, a lightweight data format is adopted. In this way, people can read and write easily; and machine parsing and

Table 4.2 Basic functional logics of human–machine interface of the automatic control system of HeXu Chinese horizontal tank

Project	Lifting pump for the collecting tank	Lifting pump for the equalization tank	Air pump	Aeration Solenoid valve	Backwash Solenoid Valve	PAC dosing pump	Carbon source dosing pump	Gas stirring valve	Electrolysis
Manual control	●	●	●	●	●	●	●	●	●
Automatic control	●	●	●	●	●	●	●	●	●
Semi-automatic commissioning	●	●	●	●	●	●	●	●	●
A pump with a backup for control	●	●	●						
Group control			●			●	●		
Level Control	●	●							
Ultra-high liquid level protection	●	●							
Standby operation			●						
Run at intervals		●	● (Standby)	●	●			●	
Alternate operation between active and standby pumps	●	●							
Active-standby failover	●	●							

(continued)

Table 4.2 (continued)

Project	Lifting pump for the collecting tank	Lifting pump for the equalization tank	Air pump	Aeration Solenoid valve	Backwash Solenoid Valve	PAC dosing pump	Carbon source dosing pump	Gas stirring valve	Electrolysis
Delay startup						●	●		
Delay stop						●	●		
Voltage regulation									●
Current regulation									●
Voltage direction regulation									●
Error alarm	●	●	●	●	●	●	●	●	●

Table 4.3 Human–machine interface functions of the automatic control system

S.N.	Functional panels	● Required ▲ Optional
1	Role-based access control	▲
2	Station information management	▲
3	Device information management	●
4	Online instrument monitoring	▲
5	Equipment data monitoring	●
6	Equipment status monitoring	●
7	Equipment parameter adjustment	●
8	Historical data view	▲
9	Logic function configuration	▲
10	Process configuration display	▲
11	Historical alarm record	●
12	Failure classification prompt	▲

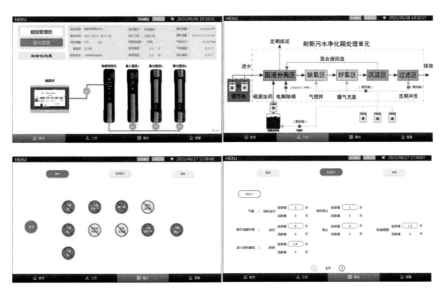

Fig. 4.6 Example of human–machine interface of the automatic control system of HeXu Chinese horizontal tank

data format generation can be easily performed, such as the most ideal general format JSON (JavaScript Object Notation).

Data variable, as a core carrier of the communication interface, communication protocol and data format, determines the specific content of data exchange between the automatic control system and the third party. It is usually set during procedure design.

4.3.3 Cloud Management System

The rural sewage treatment system is formed by a great deal of integrated sewage treatment equipment with a fixed technical process. In that case, if operation and maintenance personnel are required onsite for routine inspection and maintenance, problems such as low operation and maintenance efficiency, difficulty in operation and maintenance as well as huge costs will be caused. The cloud management and control system constructed using advanced technologies such as cloud technology, IoT, big data, and mobile Internet can remarkably enhance the operation and management efficiency of rural decentralized sewage treatment plants, reduce operation and maintenance costs, and ensure their normal and the quality of treated water. Furthermore, all work links can be seamlessly connected with the cloud management system based on the combination of PC and mobile APP, contributing to solving the problem of the previous system failing to be implemented after construction.

a. **Composition of the Cloud Management System**

The "cloud" refers to a resource pool consisting of virtual resources, storage, applications, and services. In other words, it can be interpreted as the software and hardware resources of the server being virtualized into different services, such as the software as a service (SaaS), the platform as a service (PaaS), and the infrastructure as a service (IaaS).

The information cloud management system pertains to the IaaS service model, which constitutes the equipment informatization management service platform through integrating the equipment end (control data of control unit, sensing data of online instrument, and image data of camera), the IoT end (equipment data are acquired, processed and transmitted via gateways or data acquisition instrument), and the cloud platform end (data processing service, computing service and application software service), as shown in Fig. 4.7.

The composition of the cloud management system of the integrated rural sewage treatment equipment is shown in Fig. 4.7.

Online remote management of equipment or facilities can be implemented through the establishment of a cloud management system. As can be seen from Fig. 4.8, the X1 series IoT control system equipment exchanges data with the cloud platform via the MQTT protocol, and the platform end can synchronously monitor all states and

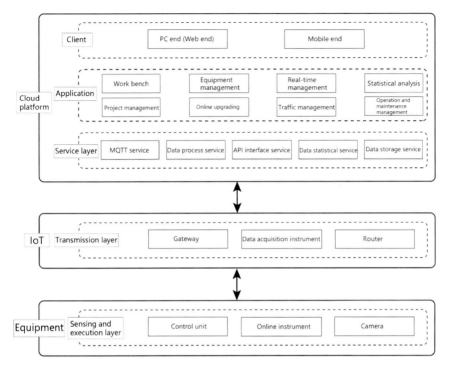

Fig. 4.7 Cloud management system framework for the integrated rural sewage treatment equipment

data of the equipment ends in real-time, realizing functions such as combined logic control, data statistical analysis, and fault management.

b. **Cloud Management System PC End (Web End)**

The PC end function of the cloud management system for the integrated rural sewage treatment equipment should be designed as per the actual usage conditions and business needs to meet the requirements of centralized management, diversified business, intuitive information, comprehensive data, simultaneous monitoring, and remote control. In general, the contents shown in Table 4.4 should be involved in the functional modules of the PC end. The web interface of the cloud management system should be also designed in line with the basic principles of simplicity, easy to use, flexible operation, clear logic, and complete functions (Fig. 4.9).

c. **Cloud Management System Mobile End (Mobile End)**

The mobile end function of the cloud management system for the integrated rural sewage treatment equipment should be designed as per the business needs and the workflow of the operation and maintenance personnel as well as in accordance with the requirements of convenience, real-time monitoring, intuitive information, timely notification, simultaneous video, quick navigation, flexibility, and high efficiency. In

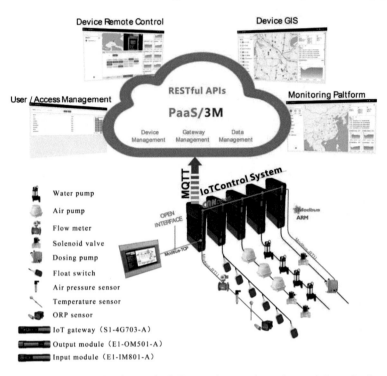

Fig. 4.8 Application scenario of X1 series IoT control system in equipment informatization cloud management

general, the contents shown in Table 4.5 should be involved in the functional modules of the mobile end of the cloud management system. The mobile end (mobile phone) interface should be also designed in accordance with the basic principles of simplicity, easy to use, flexible operation, clear logic, and complete functions. An example of the equipment commissioning interface is shown in Fig. 4.10.

Table 4.4   Functions in PC end of cloud management system of integrated rural sewage treatment equipment

S.N.	Functional panels	● Required ▲ Optional
1	Role-based access control	▲
2	Station information management	▲
3	Equipment information management	●
4	Online instrument monitoring	▲
5	Equipment data monitoring	●
6	Equipment status monitoring	●
7	Equipment parameter adjustment	●
8	Online video surveillance	▲
9	Data statistical analysis	●
10	Process configuration display	▲
11	Historical alarm record	●
12	Failure classification prompt	▲
13	Equipment online upgrading	▲
14	Project management	▲
15	Traffic management	▲
16	Operation and maintenance management	▲
17	Equipment geographic information	▲

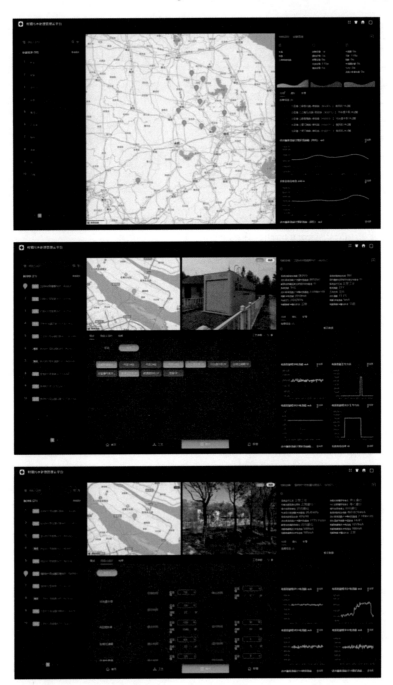

Fig. 4.9 Example of web page of the Hexu cloud management system for rural sewage treatment

Table 4.5 List of functions in mobile terminal of cloud management system of integrated rural sewage treatment equipment

S.N.	Functional panels	● Required ▲ Optional
1	Station information management	▲
2	Equipment information management	●
3	Online instrument monitoring	▲
4	Equipment data monitoring	●
5	Equipment status monitoring	●
6	Equipment parameter adjustment	●
7	Online video surveillance	▲
8	Data statistical analysis	▲
9	Process configuration display	▲
10	Historical alarm record	●
11	Failure classification prompt	▲
12	Management of operation and maintenance work order	▲
13	Equipment positioning and navigation	●

Fig. 4.10 Example of mobile terminal pages of the Hexu cloud management system for rural sewage treatment

References

Administration of Quality Supervision, Inspection and Quarantine of the People's Republic of China, Standardization Administration. (2014). *Paints and varnishes—Corrosion protection of steel structures by protective paint systems: GB/T 30790–2014*. Standards Press of China.

Construction Industry Sub-Council, China Council for the Promotion of International Trade. (2021). Assessment and certification technical specification for small-scale domestic wastewater treatment equipment: T/CCPITCUDC-002-2021. Zhongguancun Zhongke Technological Innovation and Promotion Center for Water Environment Protection.

The state bureau of Quality and Technical Supervision. (1998). *Paints and varnishes—Cross cut test for films: GB/T 9286–1998*. Standards Press of China.

Wang, H. (2018). Exploring the Chinese way of rural sewage treatment—A brief discussion on the planning, construction and management of rural sewage treatment facilities. *Water & Wastewater Engineering, 54*(5), 1–3.

Wen, Y. (2016). *Solutions for China's rural sewage treatment system*. Chemical Industry Press.

Zhang, Y., Zhou, W., & Peng, A. (2017). Survey of internet of things security. *Journal of Computer Research and Development, 54*(10), 2130–2143.

Chapter 5
Classic Cases of Integrated Rural Domestic Sewage Treatment

5.1 A Rural Domestic Sewage Treatment Case in Xiong'an New District, China

5.1.1 Project Overview

Features: Centralized Treatment on a large scale with high requirements for treated water

Construction Time: November 2019

Construction Scale: 200 m^3/d

Project objective: The quality of treated water shall conform with the discharge standard for the key control area set by the *Pollutant Discharge Standard in Daqing River Basin* (GB 18918-2018), as shown in Table 5.1.

5.1.2 Project Background

Rural sewage treatment is essential for Baiyangdian ecological environment governance according to *Baiyangdian ecological environment governance and protection planning (2018–2035)* released by Hebei, China in January 2019. This project is one of the pioneering projects for the integrated and comprehensive treatment of rural sewage, garbage, toilets, and other environmental problems in Baiyangdian, Xiong'an New Area, China (*Sewage Treatment Process Technical Route and Integrated Technology Approach for Xiong'an New Area Decentralized Villages*).

The project is located in the east of Anxin County, Baoding, Hebei, China, with an annual average temperature of 13.4 °C. The average temperatures in January and July are − 4.3 °C and 26.4 °C, respectively. With roughly 1,530 households in the

W. Li et al., *Integrated Treatment Technology of Rural Domestic Sewage*,
https://doi.org/10.1007/978-981-99-5906-8_5

Table 5.1 Engineering design for the quality of domestic sewage and treated water

Project	COD$_{cr}$	BOD$_5$	NH$_3$–N	TN	TP
Influent (mg/L)	400	200	40	50	5
Treated water (mg/L)	30	6	1.5 (2.5)	15	0.3

Note The value outside the brackets is the control index when the water temperature is greater than 12 °C while the value inside is the control index when the water temperature is smaller than or equal to 12 °C

village where the project is located, the registered population is 4,325, of which, about 1/3 are migrant workers. There are six flushing public toilets in the village, with the coverage of household toilets reaching 1/3. Daily water supply time is about 6 h, from 6:30 to 9:30 and from 16:30 to 18:00, with 1–2 h more supply time for weddings and funerals. With tap water meters installed in each household, the water charge is performed as per the actual consumption. The project serves more than 2,570 residents.

5.1.3   Treatment Scheme

Given that the project is located in an environmentally sensitive area with strict discharge standards, the improved Bardenpho-MBBR integrated sewage treatment process with outstanding denitrification and phosphorus removal was selected. The process flow is shown in Fig. 5.1, with parameters of the integrated processing equipment shown in Table 5.2. Domestic sewage is discharged into the lifting well through the collection pipe network, which then flows through the grid channel, the sand settling tank, and the equalization tank successively. After that, treated water is lifted to the integrated equipment via the lifting pump for biochemical treatment, and is discharged after reaching the standard. The excess biochemical sludge is delivered out for disposal upon being temporarily stored in the sludge tank. The ground integrated treatment equipment is selected for this project, and is covered with a thermal insulation layer to maintain the water temperature and efficiency of biochemical treatment in view of low temperature in winter. And the operating cost is 1.2 yuan/ m^3 (excluding the labor cost).

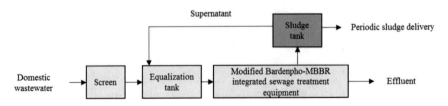

Fig. 5.1 Process flow of domestic sewage treatment in a village in Xiong'an New Area, China

Table 5.2 Parameters for modified Bardenpho-MBBR process integrated equipment

Project	Parameters	Remarks
Treatment process	Improved Bardenpho-MBBR	
Treated object	Typical domestic sewage	
Rated capacity (m^3/d)	75	
Inlet of domestic sewage	Lifted by the sewage submerged pump	
Biochemical residence time (h)	12.6	
Air demand (m^3/min)	1.03	
Type of carrier	Preferred biomass increasing sponge	
Installed power (kW)	2.88	Air pump
Operating power (kW)	1.76	Air pump
Energy consumption per ton of water ($kW{\cdot}h/m^3$)	0.56	Air pump
External material	SPA-H (Weathering steel)	White + gray
Dimension (L × W × H) (m)	12.3 × 2.4 × 2.7	
Net weight (t)	8.1	
Operating weight (t)	72.3	
Equipment installation mode	Ground type	

5.1.4 Treatment Effect

The improved Bardenpho-MBBR sewage treatment process can achieve the project objective with a satisfactory treatment effect. The treated water met the discharge limit requirements of the key control areas in the *Pollutant discharge standard of Daqing River basin* (DB13/2795-2018). Outstanding denitrification performance can be achieved with the NH_3–N of < 1 mg/L and TN of 6–10 mg/L under extremely low-temperature conditions in winter (with the water temperature of 7–8 °C). The site image is shown in Fig. 5.2.

5.1.5 Tips for Operation and Maintenance

a. The operation and maintenance management mode of "online monitoring and offline inspection" is adopted in daily operation and maintenance. Online monitoring is done remotely via the cloud platform, while offline regular inspections mainly include equipment operation and process operation.
b. A daily water quality monitoring system should be established to regularly monitor the water quality of domestic sewage and treated water.

Fig. 5.2 The site of a rural domestic sewage treatment case in Xiong'an New Area, China

c. The treatment efficiency can be improved by switching the water inlet mode by segment or adding a carbon source when the C/N ratio of domestic sewage is low.
d. The quality and quantity of domestic sewage in the project fluctuated greatly, which exerted a large impact on the treatment efficiency of the equipment. Hence, it is imperative to ensure the normal operation of the lift pump in the equalization tank and control the liquid level of the equalization tank, avoiding the long-time operation of equipment with water volume exceeding the design criteria.
e. When the devices should be shut down for a long time due to special reasons, electrical facilities such as the air pump, water pump, and electric control cabinet should be regularly maintained to ensure that they can be started normally at any time.
f. The rest should refer to the operation and maintenance tips of the improved Bardenpho-MBBR sewage treatment process (3.2.6).

5.2   A Rural Domestic Sewage Treatment Case in Kaifeng, Henan, China

5.2.1   Project Overview

Features: Household/Joint Household Treatment Mode

Construction Time: August 2019

Construction scale: Covering seven administrative villages, serving more than 5000 residents. Project objective: The quality of treated water shall be in conformity with

Table 5.3 Engineering design for the quality of domestic sewage and treated water

Project	COD_{cr}	NH_3-N	SS
Domestic sewage (mg/L)	400	60	200
Treated water (mg/L)	100	20 (25)	50

Note The value outside the brackets is the control index when the water temperature is greater than 12 °C while the value inside is the control index when the water temperature is smaller than or equal to 12 °C

the third level criteria of the *Discharge standard of pollutants for rural sewage treatment plants* (DB41/1820-2019), as shown in Table 5.3.

5.2.2 Project Background

The rural toilet revolution is highlighted in the "Three-Year Action Plan for Rural Living Environment Improvement". This was a demonstration project for the rural toilet revolution in Kaifeng, Henan, China, covering seven administrative villages, roughly 1000 households with approximately 5000 residents. The level III discharge criteria of the *Discharge standard of pollutants for rural sewage treatment plants* of Henan, China (DB41/1820-2019) was implemented. The following problems are seen in the rural domestic sewage treatment in this project:

a. Difficulty to collect domestic sewage in a centralized manner. Domestic sewage cannot be collected in a centralized manner due to the incomplete toilet construction. Since there is neither a sewage pipe nor a supporting sewage treatment plant in most of the villages, domestic sewage is usually directly discharged into nearby rivers without any treatments.
b. The sewage is incompletely collected, failing to attach importance to grey water treatment. The septic tanks in some farming households only have access to the sewage from the toilets, failing to collect the washing sewage, bathing sewage and kitchen sewage. The incomplete "four water" collection leads to low sewage treatment rate and pollution in the surrounding environment. Hence, the toilet revolution should be improved.

5.2.3 Treatment Scheme

The project is a supporting project for the toilet revolution, which is configured with a flexible selection of the single household/joint household treatment mode. And its integrated equipment has both the functions of sewage collection and treatment. High installation cost is a major problem in the single household/joint household sewage treatment. Given that installation and construction requirements are low, the mode of government organization, enterprise guidance, and farmer installation should be adopted in the project as it can effectively lower construction costs, boost farming household's participation, and promote their understanding of equipment operation, environmental remediation as well as hygiene and safety.

Multi-stage A/O biological contact oxidation integrated process equipment characterized by strong resistance to impact load was adopted in the project. The process flow of the equipment is presented as below. Domestic sewage enters the equipment through gravity flow. Then it flows through the multi-stage anoxic zone (removal of organic matters, denitrification) and multi-stage aerobic zone (filling with biomass increasing sponge carrier to remove organic matters and ammonia nitrogen) in turn after the pretreatment in the septic zone (solid–liquid separation, and anaerobic fermentation). At last, it is sedimented in the sedimentation zone. The treated water can be directly discharged or flowed into the recycled tank for irrigation. Equipment parameters are shown in Table 5.4. The equipment can be used behind a septic tank (Fig. 5.3), or in scenarios without a septic tank in combination with multiple devices. The construction cost of the project is 5000 yuan/household, and the operating cost is 0.2 yuan/m^3 (excluding the labor cost).

Table 5.4 Parameters for multi-stage A/O biological contact oxidation integrated single-tank equipment

Project	Parameters	Remarks
Treatment process	Multistage A/O biological Contact Oxidation Process	
Treated object	Domestic sewage	
Dimension (m)	Φ705 × 1550 · (D × H)	
Effective depth (m)	1.1	
Treated capacity (m^3/d)	0.4	
Type of carrier	Carrier in the anoxic zone: polyhedral hollow sphere Carrier in the aerobic zone: preferred biomass increasing sponge	
Biochemical residence time (h)	24	
Air demand (L/min)	38	
Installed power (kW)	16	
Operating power (W)	16	
Energy consumption per ton of water (kW·h/m^3)	0.96	
Net weight (kg)	25	Single tank
Installation method	Buried	

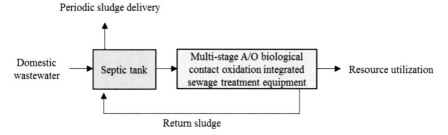

Fig. 5.3 Process flow of a rural domestic sewage treatment project in Kaifeng, Henan, China

Fig. 5.4 The site of a rural domestic sewage treatment case in Kaifeng, Henan, China

5.2.4 Treatment Effect

The project can effectively achieve the engineering objective. The integrated equipment yields favorable treatment effect, and the quality of treated water can stably reach the level III discharge criteria of the Discharge standard of water pollutants for rural sewage treatment plants of Henan, China (DB41/1820-2019). The site picture is shown in Fig. 5.4.

5.2.5 Tips for Operation and Maintenance

a. The dual mode of "daily maintenance + professional maintenance" is adopted. Farming households are responsible for routine maintenance such as domestic

Table 5.5 Engineering design for the quality of domestic sewage and treated water

Project	COD_{cr}	BOD_5	NH_3-N	TN	TP	SS
Domestic sewage (mg/L)	400	200	40	50	5	200
Treated water (mg/L)	50	10	5 (8)	15	0.5	10

Note The value outside the brackets is the control index when the water temperature is greater than 12 °C while the value inside is the control index when the water temperature is smaller than or equal to 12 °C

       sewage control, blockage treatment, and normal power supply, while professional technicians are responsible for professional maintenance such as pump cleaning, water quality monitoring, equipment commissioning, mud pumping, and equipment maintenance.

b. Operation and maintenance should be performed to ensure that facilities are in a stable power supply state for a long time due to the unique operating condition of the household facility (directly facing specific users, and the power is from the corresponding user). Otherwise, it will seriously affect the actual treatment effect of the facilities.

c. The integrated facility should be dredged regularly, in general, once a year.

d. The buried type equipment is adopted in this project. If the facility is submerged under extreme weather (such as heavy rain), the facility function will be restored through dredging after the normal weather.

5.3 A Rural Domestic Sewage Treatment Case in Liyang, Jiangsu, China

5.3.1 Project Overview

Features: Small-scale centralized treatment with high requirements for treated water

Construction time: July 2021

Construction scale: 20 m³/d

Project objective: The quality of treated water shall be in conformity with the *Pollutant discharge standard of water for municipal sewage treatment plant* (GB 18918-2002) (Table 5.5).

5.3.2 Project Background

Jiangsu is one of the earliest provinces in China to perform rural sewage treatment, with some areas achieving full coverage of rural sewage treatment. The total investment in the rural domestic sewage comprehensive treatment project in Liyang, Jiangsu, China, was estimated to be roughly 1.3 billion yuan. The project realizes comprehensive treatment of domestic sewage in 935 key villages in Liyang, Jiangsu, China, involving 1 street, 150 administrative villages, and 935 natural villages in 10 designated towns, and 91,130 beneficiary households.

The project involves the rural domestic sewage treatment in four towns, with the utilization of 155 sets of integrated treatment equipment in total, of which 110 sets of equipment conform to the treated water reaching the Level I B criteria of GB18918-2002, while 45 sets following the Level I A criteria of GB18918-2002. One station was chosen as a typical case for a detailed introduction as there is a large number of integrated facilities in the project.

The station is 18 km away from the market town with 23 natural villages (32 villager groups), covering an area of 14.7 km^2 and a total population of 3860. As most of the young people in the village work outside for a long time, the elderly, women and children are permanent residents who have a good habit to save water with a small discharge of daily sewage. But during holidays, large numbers of migrant workers returning home may lead to a surge in sewage discharge. There are also problems such as unsound construction of drainage and water collection facilities, no rain and sewage diversion, and great fluctuations in water quantity and quality with seasons and weather conditions. Its domestic sewage treatment is characterized by high-standard requirements (GB18918-2002 Grade A) for treated water, and low COD of domestic sewage with an unbalanced C/N ratio.

5.3.3   Treatment Scheme

Given the large fluctuation of water inflow and high requirements for the quality of treated water on site, a multi-stage A/O biological contact oxidation sewage integrated treatment process was selected with strong resistance to impact load and outstanding treatment performance. An electrolytic dephosphorization module was added to the integrated equipment for its low efficiency in dephosphorization. The project process is shown in Fig. 5.5, with parameters of the integrated treatment equipment shown in Table 5.6. Its main structure is a lifting well with the dimension of 1000 * 2000 * 4000 mm. The specific process is presented below. First, the domestic sewage is collected into the lifting well via the household oil separating tank and septic tank. Then, suspended or precipitated super-large solid matters are separated by the grid. On this basis, sewage flows into the flow equalization tank and is pumped to the multi-stage A/O biological contact oxidation equipment by the lifting pump, and is discharged after meeting the standard. And the sludge produced by the system is transported regularly. Its operating cost is 1.4 yuan/m^3 (excluding the labor cost).

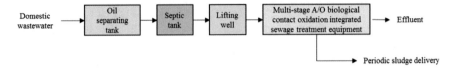

Fig. 5.5 Process flow of a rural domestic sewage treatment project in Liyang, Jiangsu, China

Table 5.6 Parameters for multi-stage A/O biological contact oxidation process integrated equipment

Project	Parameters	Remarks
Treatment process	Multi-stage A/O biological contact oxidation process	
Treated object	Rural domestic sewage	
Inlet of domestic sewage	Intermittent feeding by the sewage submerged pump	
Biochemical residence time (h)	32	
Air demand (m^3/min)	0.25	
Type of carrier	Preferred biomass increasing sponge	
Installed power (kW)	1.56	Air pump
Operating power (kW)	0.34	Air pump
Energy consumption per ton of water ($kW \cdot h/m^3$)	1.63	Air pump
Device external material	PP + glass fiber	Light grey
Dimension (L × W × H) (m)	5.5 × 2.0 × 2.3	
Net weight (t)	1.36	
Operating weight (t)	11.7	
Equipment installation mode	Buried	

5.3.4   Treatment Effect

The multi-stage A/O biological contact oxidation process for sewage treatment can achieve the engineering target effectively. The project equipment is stably operated with a satisfactory treatment effect. The quality of treated water can meet the Level I A criteria of the *Pollutant Discharge Standard of Water for Municipal sewage treatment plants* (GB18918-2002). The site picture is shown in Fig. 5.6.

Fig. 5.6   The site of a rural domestic sewage treatment case in Liyang, Jiangsu, China

5.3.5   Tips for Operation and Maintenance

a. The operation and maintenance management mode of "online monitoring and offline inspection" is adopted in daily operation and maintenance. Online monitoring is done remotely via the cloud platform, while offline regular inspections mainly include equipment operation (especially the electrolysis dephosphorization module) and process operation.
b. A daily water quality monitoring system should be established to regularly monitor the water quality of domestic sewage and treated water.
c. To ensure the normal operation of equipment, the on-site sludge should be cleaned regularly. In general, the interval for extracting sludge for delivery should be every three to six months. Besides, the sludge accumulation of the equipment should be observed regularly.
d. The buried type equipment is adopted in this project. If the facility is submerged under extreme weather (such as heavy rain), the facility function will be restored through dredging after the normal weather.
e. Oil pollution of the oil separation facility in the pretreatment part of the facility should be cleaned timely to ensure that oil pollution is prevented from entering the biochemical process section of the facility. In general, oil separation facilities should be cleaned once every 1–2 months.
f. Carbon sources should be added regularly because of the low COD and unbalanced C/N ratio of the domestic sewage on-site.
g. The rest should refer to the operation and maintenance tips of the A/O biological contact oxidation treatment process (3.4.6).

Table 5.7 Engineering design for the quality of domestic sewage and treated water

Project	COD_{cr}	$NH_3–N$	TN	TP	SS
Domestic sewage (mg/L)	400	40	50	6	200
Treated water (mg/L)	60	8 (15)	30	3	20

Note The value outside the brackets is the control index when the water temperature is greater than 12 °C while the value inside is the control index when the water temperature is smaller than or equal to 12 °C

5.4   A Rural Domestic Sewage Treatment Case in Nanjing, Jiangsu, China

5.4.1   Project Overview

Features: Small-scale centralized treatment with low discharge standard for N/P

Construction time: July 2019

Construction scale: 10 m³/d

Project objective: The quality of treated water shall conform with the Level I B criteria of the *Discharge standard of water pollutants for rural sewage treatment plants* (DB32 3462-2020), as shown in Table 5.7.

5.4.2   Project Background

The project is located in Liuhe District, Nanjing, Jiangsu, China. As it was implemented after the issuance of the *Water pollutant discharge standard for rural domestic sewage treatment plants* (DB32 3462-2020), the standard described in Sect. 5.3 is no longer implemented. Pilot projects were made in 711 rural living environment improvement demonstration villages such as Shanbei Community of Xiongzhou Street and Shuangdun Community of Yeshan Street to complete the full coverage of sewage treatment plants in all-natural villages in Liuhe District, according to the *"Proposals for Implementing Rural Living Environment Improvement Demonstration Village Construction of Rural Sewage Treatment plants in Liuhe District in 2019".*

This project undertakes the rural domestic sewage treatment plants in six towns and 412 stations. One site was selected as a typical case for a detailed introduction considering the extensive design scope of the project and the large number of integrated facilities. The project serves 67 households with a registered population of 218, including 144 permanent residents. Its domestic sewage treatment is characterized by relaxed requirements for TN and TP treatment, high requirements for removal of ammonia nitrogen, and low COD of domestic sewage with an unbalanced C/N ratio.

5.4.3 Treatment Scheme

Given the large fluctuation of water inflow and low requirements for the treatment of TN, a SND biological contact oxidation integrated sewage treatment process was selected with strong resistance to impact load and a simplified process. An electrolytic dephosphorization module was added to the integrated equipment for considering its low efficiency in dephosphorization. The process flow of the project is shown in Fig. 5.7.

Parameters for the integrated treatment equipment are shown in Table 5.8. Domestic sewage collected by the pipeline flows into the equalization tank of the centralized sewage treatment station after entering the septic tank and the oil separating tank, which is then lifted by the equalization tank pump to the integrated treatment equipment for treating and is discharged after meeting the standard. The operating cost is 1.07 yuan/m^3 (excluding the labor cost).

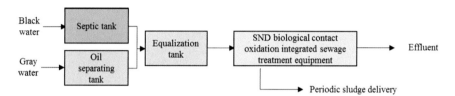

Fig. 5.7 Process flow of a rural domestic sewage treatment project in Nanjing, Jiangsu, China

Table 5.8 SND biological contact oxidation process integrated equipment parameters

Project	Parameters	Remarks
Treatment process	SND type biological contact oxidation process	
Treated object	Typical domestic sewage	
Inlet of domestic sewage	Lifted by the sewage submerged pump	
Biochemical residence time (h)	13	
Air demand (m^3/min)	0.4	
Type of carrier	Preferred biomass increasing sponge	
Installed power (kW)	0.43	Air pump
Operating power (kW)	0.36	Air pump
Energy consumption per ton of water (kW·h/m^3)	0.86	Air pump
External material	PPH (PP material)	White or cream
Dimension (L × W × H) (m)	5.0 × 2.0 × 3.0	
Net weight (t)	1.1	
Operating weight (t)	13.9	
Equipment installation mode	Buried	

Fig. 5.8 The site of a rural domestic sewage treatment case in Liuhe district, Nanjing, China

5.4.4   Treatment Effect

Simultaneous nitrification and denitrification (SND) process adopted can achieve the project objective effectively. The integrated equipment is stably operated with a satisfactory treatment effect. The quality of treated water can meet the Level I B criteria of the *Discharge standard of water pollutants for rural sewage treatment plants of Jiangsu, China* (DB32 3462-2020). The site picture is shown in Fig. 5.8.

5.4.5   Tips for Operation and Maintenance

a. The operation and maintenance management mode of "online monitoring and offline inspection" is adopted in daily operation and maintenance. Online monitoring is done remotely via the cloud platform, while offline regular inspections mainly include equipment operation (especially the electrolysis dephosphorization module) and process operation.
b. A daily water quality monitoring system should be established to regularly monitor the water quality of domestic sewage and treated water.
c. To ensure the normal operation of equipment, the on-site sludge should be cleaned regularly. In general, the interval for extracting sludge for delivery should be every 3 to 6 months. Besides, the sludge accumulation of the equipment should be observed regularly.
d. The buried equipment is adopted in this project. If the facility is submerged under extreme weather (such as heavy rain), the facility function will be restored through dredging after the normal weather.
e. The rest should refer to the operation and maintenance tips of the SND biological contact oxidation process.

Table 5.9 Engineering design for the quality of domestic sewage and treated water

Project	COD_{cr}	NH_3-N	TN	TP	SS
Domestic sewage (mg/L)	80	30	35	3	50
Treated water (mg/L)	50	8	15	1	10

5.5   A Rural Domestic Sewage Discharge Standard Upgrading Project in Chongming District, Shanghai, China

5.5.1   Project Overview

Features: Tail water treatment with high requirements for treated water

Construction time: July 2021

Construction scale: 30 m³/d

Project objective: The quality of treated water shall be in conformity with the Level I A criteria of the *Discharge Standard of Water Pollutants for Rural Sewage Treatment plants* (DB31T 1163-2019), as shown in Table 5.9.

5.5.2   Project Background

32 rural domestic sewage treatment projects were conducted from 2017 to 2018 in Chongming District, Shanghai, China, which involves 18 towns and 234,000 households, with 17,700 treatment plants constructed. Japanese johkasou was adopted as sewage treatment equipment. The Shanghai Water Affairs Bureau found that the nitrogen and phosphorus concentration of the treated water from the johkasou failed to meet the Level I A criteria of the *Discharge Standard of Pollutants of Rural Domestic sewage Treatment plants* (DB31T 1163-2019) in Shanghai in a survey on the preliminary rural sewage treatment project in 2020. In response to the above problem, the whole-area upgrading and reconstruction project was proposed with a requirement that domestic sewage or johkasou treated water in areas with the condition for laying pipelines should be collected through the pressure tubes for unified treatment. For those with no condition to lay pipes, johkasou drains should be combined with the new establishment of back-end treatment plants to improve the quality of tailwater. Roughly 110,000 households are involved in the modification, and 1000 new sewage treatment plants should be built.

Construction tasks of 40 rural domestic sewage tailwater upgrading sites (with the treatment capacity of 20–120 m³/d) in three towns in Chongming District, Shanghai, China were undertaken in this project. One of the sites with a treated capacity of 30

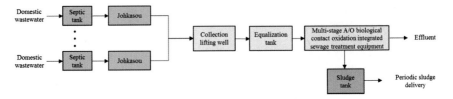

Fig. 5.9 Process flow of a rural domestic sewage discharge standard upgrading project in Chongming, Shanghai, China

m³/d was taken as an example. Domestic sewage of the station features: (a) NH₃–N, TN and TP failing to reach the Level I A criteria of DB31T 1163-2019; (b) low COD leading to the unbalanced C/N ratio; (c) large variations in inlet volume, that is, the inflow volume is too small to reach the designed treatment capacity at normal times, whereas the inflow volume is rather large at holidays.

5.5.3   Treatment Scheme

Given the low COD concentration of water inflow and high requirements for the quality of treated water, a multi-stage A/O biological contact oxidation sewage integrated treatment process was selected with strong resistance to impact load, outstanding treatment performance, and less input of carbon sources. An electrolytic dephosphorization module was added to the integrated equipment considering its low efficiency in dephosphorization. The project process flow is shown in Fig. 5.9. Parameters for the integrated processing equipment are shown in Table 5.10. The main process of sewage treatment is presented as following. Domestic sewage collected from each household first flows into the johkasou for treatment after entering the septic tank, with the treated tail water collected through the lifting well. Then, sewage enters the equalization tank in a centralized manner under pump power. After that, it is lifted into the integrated treatment equipment for end treatment by the pump in the equalization tank, which is finally discharged up to the standard. The operating cost is 0.78 yuan/m³ (excluding the labor cost).

5.5.4   Treatment Effect

The multi-stage A/O biological contact oxidation process can achieve the engineering target effectively. The integrated treatment equipment is stably operated with a satisfactory treatment effect. The quality of treated water can be in conformity with the Level I B criteria of the *Discharge Standard of Water Pollutants for Rural Sewage Treatment plants* of Shanghai, China (DB31T 1163-2019). The site picture is shown in Fig. 5.10.

Table 5.10 Multi-stage A/O biological contact oxidation process integrated equipment parameters

Project	Parameters	Remarks
Treatment process	Multi-stage A/O biological contact oxidation process	
Treated object	Advanced treatment of tail water of the johkasou	
Rated capacity (m³/d)	30	
Inlet of domestic wastewater	Lifted by the sewage submerged pump	
Biochemical residence time (h)	16	
Air demand (m³/min)	1.0	
Type of carrier	Anaerobic efficient carrier + biomass ascending carrier	
Installed power (kW)	1.34	Air pump
Operating power (kW)	1.09	Air pump
Energy consumption per ton of water (kW·h/m³)	0.87	Air pump
External material	SPA-H (Weathering steel)	White + gray
Dimension (L × W × H) (m)	6.5 × 2.4 × 2.7	
Net weight (t)	6.0	
Operating weight (t)	40	
Equipment installation mode	Ground type	

Fig. 5.10 The site of a rural domestic sewage discharge standard upgrading case in Chongming District, Shanghai, China

5.6   Tips for Operation and Maintenance

a. The operation and maintenance management mode of "online monitoring and offline inspection" is adopted in daily operation and maintenance. Online monitoring is done remotely via the cloud platform, while offline regular inspections mainly include equipment operation (especially the electrolysis dephosphorization module) and process operation.
b. A daily water quality monitoring system should be established to regularly monitor the water quality of domestic sewage and treated water.
c. Carbon sources should be supplemented to get the TN of treated water under control. The supplementary amount of carbon source is determined by the C/N ratio of domestic sewage and the actual removal efficiency of TN.
d. Process operation parameters can be flexibly controlled according to the seasonal characteristics of the quality of tail water from the johkasou. If the quality of tail water is good in summer, the aeration volume of the aerobic zone and the reflux ratio of nitrifying liquid can be appropriately lower to achieve energy saving and carbon reduction.
e. To ensure the normal operation of equipment, the on-site sludge should be cleaned regularly. In general, the interval for extracting sludge for delivery is done every 3 to 6 months. Besides, the sludge accumulation of the equipment should be observed regularly.
f. The cleaning of sundries such as leaves, and branches should be highlighted during maintenance since the equipment station is set up in the forest. It is aimed at preventing these sundries from falling into the equipment, thereby affecting the treatment effect of the equipment.
g. The rest refers to the operation and maintenance points of the multi-stage A/O biological contact oxidation process (3.4.6).

5.7   A Rural Domestic Sewage Treatment Case in Huangshan, Anhui, China

5.7.1   Project Overview

Features: Centralized Treatment in Large Scale

Construction Time: September 2019

Construction Scale: 200 m^3/d

Project objective: The quality of treated water shall be in conformity with the first level B criteria of the *Discharge Standard of pollutants for Municipal Sewage Treatment Plants* (GB 18918-2002), as shown in Table 5.11.

Table 5.11 Engineering design for the quality of domestic sewage and treated water

Project	COD_{cr}	BOD_5	NH_3–N	TN	TP	SS
Domestic sewage (mg/L)	400	200	40	50	5	200
Treated water (mg/L)	60	20	8 (15)	20	1	20

Note The value outside the brackets is the control index when the water temperature is greater than 12 °C while the value inside is the control index when the water temperature is smaller than or equal to 12 °C

5.7.2 Project Background

The Huangshan Municipal People's Government decided to implement the PPP project of rural sewage treatment to improve the rural sewage treatment capacity, guarantee the quality of the water environment, and ensure the safety of drinking water in the downstream areas and the Yangtze River Delta region. The PPP project is expected to operate for 15 years with a total investment of 450,000,000 yuan, and 70 plant stations were built with a treatment capacity of 8000 m^3/d. This project is part of the PPP phase 1 project for rural sewage treatment in Huangshan, Anhui, China, which serves 1 village with 595 households covering 2055 people, including 1678 permanent residents. Domestic sewage discharge is characterized by:

(a) Residents living in a concentrated area are convenient for the centralized collection and treatment of domestic sewage; (b) the standard of treated water is lax.

5.7.3 Treatment Scheme

Given the lax standard for water treatment of the project and the adoption of a large-scale centralized treatment mode, the A^3/O-MBBR (mud-film composite) integrated sewage treatment process with small floor space, and stable and low energy consumption is selected in this project. The process flow is shown in Fig. 5.11, with parameters of the integrated processing equipment shown in Table 5.12. The main process is presented below: domestic sewage collected by the pipeline flows into the equalization tank by gravity for homogeneous and equalized treatment after large, suspended solids are removed by the screen. Then, the sewage is pumped into the A^3/O-MBBR (mud-film composite) integrated sewage treatment equipment and is discharged after reaching the standard. And the excess sludge is discharged into the sludge thickening tank before delivering out for landfill or composting. The integrated equipment for sewage treatment is installed on the ground. The construction cost of the project is 5,000 yuan/m^3, and the operating cost is 0.7 yuan/m^3 (excluding the labor cost).

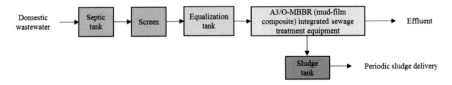

Fig. 5.11   Process flow of a rural domestic sewage treatment project in Huangshan, Anhui, China

Table 5.12   Parameters for A³/O-MBBR process integrated equipment

Project	Parameters	Remarks
Treatment process	A³/O-MBBR process	Mud-film composite
Treated object	Typical domestic sewage	
Rated capacity for water treatment (m³/d)	100	
Inlet of domestic sewage	Lifted by the sewage submerged pump	
Biochemical residence time (h)	10.4	
Air demand (m³/min)	1.17	
Type of carrier	Preferred biomass increasing sponge	
Installed power (kW)	2.32	
Operating power (kW)	1.72	
Energy consumption per ton of water (kW·h/m³)	0.41	
External material	SPA-H (Weathering steel)	White + gray
Dimension (L × W × H) (m)	12.3 × 2.4 × 2.7	
Net weight (t)	9.2	
Operating weight (t)	68.7	
Equipment installation mode	Ground type	

5.7.4   Treatment Effect

A³/O-MBBR integrated sewage treatment process can achieve the project objective effectively. The integrated treatment equipment is operated in good condition. The quality of treated water is in conformity with the first level B criteria of the "*Pollutant discharge standard for municipal sewage treatment plants*" (GB 18918-2002). The site picture is shown in Fig. 5.12.

Fig. 5.12 The site of a rural domestic sewage treatment case in Huangshan, Anhui, China

5.7.5 Operation and Maintenance

a. The operation and maintenance management mode of "online monitoring and offline inspection" is adopted in daily operation and maintenance. Online monitoring is done remotely via the cloud platform, while offline regular inspections mainly include equipment operation and process operation.
b. A daily water quality monitoring system is established to regularly monitor the water quality of domestic sewage and treated water.
c. Daily sludge discharge of the equipment is reasonably adjusted according to the water quality and quantity.
d. The rest refer to the operation and maintenance points of the A^3/O-MBBR process (3.1.6)

5.8 A Domestic Sewage Treatment Case in Chongqing Yangtze River Protection Rural Scenic Spots

5.8.1 Project Overview

Features: Centralized Treatment in Large Scale

Construction Time: June 2021

Construction Scale: 250 m^3/d

Table 5.13 Engineering design for the quality of domestic sewage and treated water

Project	COD_{cr}	NH_3-N	TP	SS
Domestic sewage (mg/L)	400	40	5	200
Treated water (mg/L)	80	20	3	30

Project objective: The quality of treated water shall be in conformity with the first level of the *Discharge standard of water pollutants for rural sewage treatment plants* (GB 18918-2018), as shown in Table 5.13.

5.8.2   Project Background

As the Yangtze River is the mother river of the Chinese nation, its ecological protection and restoration are highly concerned with sustainable economic and social development. The Yangtze River Protection Law of the People's Republic of China came into force on 1st March 2021. And Chongqing plays an irreplaceable role in the ecological security of the middle and lower reaches of the Yangtze River as the last pass of the ecological barrier in the upper reaches. 16 new sewage treatment stations in Jiangjin District and 11 new sewage treatment stations in Liangping District are involved in the Chongqing Yangtze River Protection Sewage Treatment Project, covering integrated equipment with treatment capacities of 20 m^3/d (2 sets), 30 m^3/d (1 set), 40 m^3/d (3 sets), 50 m^3/d (28 sets), 60 m^3/d (2 sets), 75 m^3/d (8 sets), 120m^3/d (4 sets), 125 m^3/d (2 sets) set), and 150 m^3/d (2 sets) in total. One of the stations covering 120 m^3/d is undertaken in this project. The sewage at this site is from the scenic spot, featuring that (a) There are tourist distribution points such as farmhouses and summer resorts in the scenic spot. (b) The complete sewage pipe network in the scenic spot can collect sewage easily. In this way, sewage can be treated in a large centralized mode. (c) Sewage volume varies widely in different seasons. In winter, there is a small sewage volume for there are only permanent residents in the scenic spot. But there are many tourists in summer with 400–800 tourists per day. The water consumption increases together with the rising sewage discharge.

5.8.3   Treatment Scheme

The multi-stage A/O contact oxidation process with a strong resistance to impact load is adopted considering that the scenic spot where the project is located has a great difference in water quality and quantity. The process flow is shown in Fig. 5.13, with parameters of the integrated processing equipment shown in Table 5.14. The domestic sewage flows into the equalization tank through the screen and is discharged up to the standard after being treated by the multi-stage A/O biological contact oxidation

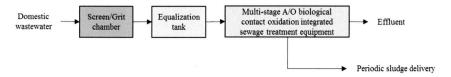

Fig. 5.13 Process flow of a sewage treatment case of Yangtze River protection in Chongqing, China

Table 5.14 Parameters for multi-stage A/O biological contact oxidation process integrated treatment equipment

Project	Parameters	Remarks
Treatment process	Multi-stage A/O biological contact oxidation process	
Treated object	Rural domestic sewage	
Inlet of domestic sewage	Lifted by the sewage submerged pump	
Biochemical residence time (h)	16	
Air demand (m^3/min)	1.0	
Type of carrier	Anaerobic efficient carrier + biomass ascending carrier	
Installed power (kW)	1.34	Air pump
Operating power (kW)	1.09	Air pump
Energy consumption per ton of water ($kW{\cdot}h/m^3$)	0.87	Air pump
Device external material	SPA-H (Weathering steel)	Light gray
Dimension (L × W × H) (m)	6.5 × 2.4 × 2.7	Signal box
Net weight (t)	6.0	
Operating weight (t)	40	
Equipment installation mode	Buried	

integrated treatment equipment. The integrated equipment for sewage treatment is installed on the ground, with the operating cost of 1.1 yuan/m^3 (Excluding the labor cost).

5.8.4   Treatment Effect

The multi-stage A/O biological contact oxidation process for sewage treatment can achieve the engineering objective effectively. The integrated treatment equipment is operated in good condition. The quality of treated water is in conformity with the first level criteria of the "*Pollutant discharge standard for rural sewage treatment plants*" (DB50/848–2018).

The project site picture is shown in Fig. 5.14.

Fig. 5.14 The site of a rural domestic sewage treatment case in Chongqing Yangtze River protection spots

5.8.5   Operation and Maintenance

a. The operation and maintenance management mode of "online monitoring and offline inspection" is adopted in daily operation and maintenance. Online monitoring is done remotely via the cloud platform, while offline regular inspections mainly include equipment operation and process operation.

b. A daily water quality monitoring system is established to regularly monitor the water quality of domestic sewage and treated water.

c. As seasonal differences in scenic spots are large, inspections should be strengthened during the peak tourist seasons, with operational monitoring records taken.

d. To ensure the normal operation of equipment, the on-site sludge should be cleaned regularly. In general, the interval for extracting sludge for delivery should be every 3 to 6 months. Besides, the sludge accumulation of the equipment should be observed regularly.

e. The rest should refer to the operation and maintenance tips of the multi-stage A/O biological contact oxidation process (3.4.6).

Table 5.15 Reference of the quality of domestic sewage and treated water of the project (unit: mg/L)

Project	COD_{Cr}	BOD_5	NH_3-N	TN	TP	SS
Designed quality of domestic sewage	400	200	40	50	5	200
Designed quality of treated water	50	10	5 (8)	15	0.5	10

Note The value outside the brackets is the control index when the water temperature is greater than 12 °C while the value inside is the control index when the water temperature is smaller than or equal to 12 °C

5.9 A Rural Domestic Sewage Treatment Case in Shiping County, Yunnan, China

5.9.1 Project Overview

Features: Decentralized + centralized, high requirements for treated water

Construction time: April 2017

Construction scale: 10 m^3/d

Project objective: The quality of treated water shall be in conformity with the Level I A criteria of the *Pollutant Discharge Standard of Water for Municipal sewage treatment plants* (GB 18918-2002), as shown in Table 5.15.

5.9.2 Project Background

"Twelfth Five-Year Plan" for the comprehensive prevention and control of water pollution in the Yilong Lake Basin was formulated by the Shiping County, Yunnan Province in response to the *"Guiding Opinions on Improving Rural Living Environment"* (2014) issued by State Council. Yilong Lake, located in the southeast of Shiping County, is one of the nine plateau lakes in Yunnan. The Yilong Lake Basin is the area with the densest population in Shiping County. The irregular sewage discharge by villages in the basin significantly polluted lake water and deteriorated the aquatic ecological environment. Hence, a comprehensive environmental improvement project for villages along the Yilong Lake was planned by the government. There are two types of sewage treatment methods according to the project requirements. The first is centralized treatment, the water quality after sewage treatment reaches the Level I A standard of *Pollutant Discharge Standard of Water for Municipal sewage treatment plants* (GB18918-2002). The other is household treatment, the water quality upon the sewage treatment reaches Level I B criteria of GB18918-2002 (except TP).

Seventy-three centralized facilities and 887 household sewage treatment plants are constructed in this project. A^3/O-MBBR process and its integrated equipment and multi-stage A/O process and its integrated equipment are utilized in centralized sewage treatment (with the treatment capacity of 10–200 m^3/d). SND process and its integrated equipment are adopted in household sewage treatment (with a treatment capacity of 0.6–5 m^3/d). The sewage treatment project started in 2017, and three batches of equipment supply, installation, and commissioning were implemented at the end of October 2017, 2018, and 2019 respectively.

The construction of one of the small centralized processing sites was involved in this project. It serves more than 50 farmer households, with more than 150 permanent residents. Its domestic sewage discharge is characterized by: (a) Project construction such as sewage pipelines, garbage collection rooms, public toilets, and composting tanks was completed, except for sewage treatment. (b) Sewage can be collected and treated in a centralized manner. (c) The ground installation can be selected if the temperature difference between the four seasons is small with no extreme low temperature in winter.

5.9.3   Treatment Scheme

Given the large fluctuation of water inflow and high requirements for the quality of treated water on site, a multi-stage A/O biological contact oxidation sewage integrated treatment process was selected with strong resistance to impact load and outstanding treatment performance. The process flow is shown in Fig. 5.15, with parameters of the integrated equipment shown in Table 5.16. First, the domestic sewage is collected into the lifting well via the household oil separating tank and septic tank, with suspended or precipitated super-large solid matters separated by the grid. On this basis, sewage is pumped to the multi-stage A/O biological contact oxidation integrated treatment equipment for treating and discharging after meeting the standard. And the sludge produced by the system is transported regularly. The integrated equipment for sewage treatment is installed on the ground, with an operating cost of 0.5 yuan/m^3 (excluding the labor cost).

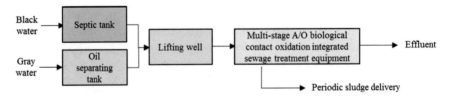

Fig. 5.15 Process flow of a rural sewage treatment case in Shiping County, Yunnan, China

Table 5.16   Parameters for multi-stage A/O biological contact oxidation process equipment

	Parameters	Remarks
Treatment process	Multi-stage A/O biological contact oxidation process	
Treated object	Domestic sewage	
Inlet of domestic sewage	Intermittent water inlet	
Air demand (m^3/min)	0.5	
Type of carrier	Polyhedral hollow sphere + biomass increasing sponge	
Installed power (kW)	1.21	Air pump
Operating power (kW)	0.4	Air pump
Energy consumption per ton of water ($kW \cdot h/m^3$)	0.96	Air pump
External material	Q235-B	
Dimension (L × W × H) (m)	6.09 × 1.31 × 2.2	
Net weight (t)	2	
Operating weight (t)	9	
Equipment installation mode	Ground type	

5.9.4   Treatment Effect

The multi-stage A/O biological contact oxidation process can achieve the engineering objective effectively. The project is stably operated with a satisfactory treatment effect. The quality of treated water can meet the Level I A criteria of the *Pollutant Discharge Standard of Water for Municipal sewage treatment plants* (GB18918-2002). The site picture is shown in Fig. 5.16.

5.9.5   Tips for Operation and Maintenance

a. Regular inspection, including equipment and process operation should be performed. Dissolved oxygen in the aerobic zone should be regularly checked to ensure it stays within a reasonable range. The domestic sewage inlet and the aeration of each aerobic functional zone should be calibrated according to the influent load.
b. A daily water quality monitoring system should be established to regularly monitor the water quality of domestic sewage and treated water.
c. To ensure the normal operation of equipment, the on-site sludge should be cleaned regularly. In general, the interval for extracting sludge for delivery should be every 3 to 6 months. Besides, the sludge accumulation of the equipment should be observed regularly.

Fig. 5.16 Site of a rural sewage treatment case in Shiping County, Yunnan, China

d. The rest should refer to the operation and maintenance tips of the multi-stage A/
   O biological contact oxidation process (3.4.6).

5.10   A Rural Domestic Sewage Emergency Treatment Case in Dongguan, Guangdong, China

5.10.1   Project Overview

Features: Centralized and emergency treatment on large scale

Construction Time: January 2020

Construction Scale: 8000 m³/d

Project objective: The quality of treated water shall be in conformity with the Level
I A criteria of *the Pollutant Discharge Standard of Water for Municipal sewage
treatment plants* (GB 18918-2002), as shown in Table 5.17.

Table 5.17 Engineering design for the quality of domestic sewage and treated water

Project	COD_{cr}	BOD_5	NH_3-N	TN	TP	SS
Domestic sewage (mg/L)	400	200	40	50	5	200
Treated water (mg/L)	50	10	5 (8)	15	0.5	10

Note The value outside the brackets is the control index when the water temperature is greater than 12 °C while the value inside is the control index when the water temperature is smaller than or equal to 12 °C

5.10.2 Project Background

There emerged problem of the amount of sewage collected in the township far exceeding the design scale of the sewage treatment plant in Dongkeng Town, Dongguan, Guangdong, China in the construction of secondary sewage interception pipeline network. Meanwhile, the existing sewage treatment plant is expanding and upgrading. Consequently, the sewage in some areas of Dongkeng Town could not be transported to the sewage treatment plant for treatment, resulting in the river being seriously polluted as well as black and odorous water. To this end, an emergency treatment project was performed. The project features (a) Large volume and high discharge standard. With the daily treatment capacity of 8000 m³, the quality of the treated water shall meet the Level I A criteria of the *Pollutant Discharge Standard of Water for Municipal sewage treatment plants* (GB 18918-2002). (b) Emergency treatment with a short construction period. The designed construction period from the design implementation to the realization of discharging after meeting the standard is only 40 days.

5.10.3 Treatment Scheme

The A/O-MBBR integrated sewage treatment process with a simple process, low investment and low operating cost was selected by the project. The technological process is shown in Fig. 5.17. The composition of the main structure of the project is shown in Table 5.18, and the equipment parameters are shown in Table 5.19. With the total designed treatment capacity of 8000 m³/d, eight sets of integrated treatment systems were connected in parallel, with each formed by three integrated treatment plants connected in series. Considering the low efficiency in phosphorus removal, the A/O-MBBR integrated sewage treatment process is equipped with a chemical phosphorus removal module (PAC). The main technological process is presented as following: Domestic sewage is discharged into the integrated lifting pump station through the collection pipe network and flows through the screen channel, the sand settling tank and the equalization tank successively. Then the influent is distributed to the integrated sewage treatment equipment by the distribution well and deeply treated via the cloth filter. Finally, the treated water is discharged or recycled after

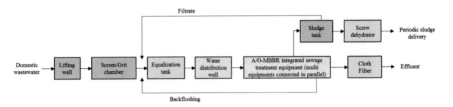

Fig. 5.17 Process flow of the rural domestic sewage emergency treatment case in Dongguan, Guangdong, China

Table 5.18 Main structures of the rural domestic sewage emergency treatment case in Dongguan, Guangdong, China

Name	Dimension (L × W × H) (m)	Structure
Fine screen ditch	11 × 3.9 × 1.6	RC
Equalization tank	36.3 × 6 × 4.5	RC
Disinfection discharge ditch	1.3 × 6 × 1.3	RC
Sludge tank	6 × 2 × 4.5	RC
Online monitoring room	4.5 × 4.5 × 3.5	Prefabricated house
Dosing Room	8 × 4.5 × 3.5	Prefabricated house
Blower room	9.5 × 45 × 3.5	Prefabricated house
Sludge dewatering room	10.5 × 4.5 × 3.5	Prefabricated house
Central control room	4 × 4.5 × 3.5	Prefabricated house
Office	–	RC

being disinfected and sterilized. The excess biochemical sludge being temporarily stored in the sludge tank is filtered by the screw press before it is transported out for disposal. The integrated equipment for sewage treatment is installed on the ground, with an operating cost of 1.01 yuan/m^3 (excluding the labor cost).

5.10.4 Treatment Effect

The A/O-MBBR sewage treatment process can achieve the emergency treatment objective effectively. The project was completed as scheduled, with the installation done in the 5th week, and the discharge after meeting standards in the 6th week. The integrated equipment was operated stably with a satisfactory treatment effect. And all indicators of the quality of the treated water can steadily meet the discharge standard.

The site picture is shown in Fig. 5.18.

Table 5.19 Parameters of A/O-MBBR integrated treatment equipment

Project	Parameters	Remarks
Treatment process	A/O-MBBR	
Treated object	Typical domestic sewage	
Rated capacity for sewage treatment (m³/d)	1000	A single system
Inlet of domestic sewage	Lifted by the sewage submerged pump	
Biochemical residence time (h)	7.2	
Air demand (m³/min)	10.5	
Type of carrier	Preferred biomass increasing sponge	
Installed power (kW)	4.2	Fan
Operating power (kW)	3.5	Fan
Energy consumption per ton of water (kW·h/m³)	0.43	Fan
External material	SPA-H (Weathering steel)	Light grey + grey
Dimension (L × W × H) (m)	17.5 × 2.9 × 2.9	Single equipment
Net weight (t)	13.1	
Operating weight (t)	72.3	
Equipment installation mode	Ground type	

Fig. 5.18 The site of rural domestic sewage emergency treatment case in Dongguan, Guangdong, China

5.10.5 Tips for Operation and Maintenance

a. The operation and maintenance contain eight sets of integrated treatment systems, and other auxiliary facilities supporting the same station, such as a cyclone settling tank, mechanical screen, water collection tank, lifting pump, and water distribution tank.
b. Personnel specializing in operation, maintenance, and repair should be arranged to regularly inspect the operation of equipment and facilities and monitor the quality of domestic sewage and treated water.
c. Sewage in zone A should be uniformly mixed to avoid local sludge accumulation. Timely dredging is needed if the mixture pipeline is found blocked.
d. Carriers in the O zone should be checked for no accumulation, no damage, and no non-functional biological overload.
e. The carrier block at the tail end of the O zone should be kept unobstructed.
f. The sludge accumulated on the inclined tube carrier in the sedimentation zone should be backwashed in time.
g. The filtration unit of the filter tank should be cleaned regularly.
h. As fan aeration is adopted, the condition of the fan belt should be checked daily to find abnormalities such as oil leakage and abnormal noise since. Also, the filter screen of the fan air inlet must be cleaned regularly.
i. Whether the dosing unit is normally operated should be checked regularly to avoid blockage, empty pumping, or insufficient dosage.
j. The feeding and dosing amounts can be adjusted when abnormal dehydration is found in the screw dehydrator. Meanwhile, the blockage should be removed in time.

5.11 A Campus Domestic Sewage Treatment Case in Haikou, Hainan, China

5.11.1 Project Overview

Features: Centralized treatment on a large scale with high requirements for treated water

Construction time: December 2020

Construction scale: 1500 m³/d

Project objective: The quality of treated water shall be in conformity with the Level I A criteria of the *Pollutant Discharge Standard of Water for Municipal sewage treatment plants* (GB 18918-2002), as shown in Table 5.20.

Table 5.20   Engineering design for the quality of domestic sewage and treated water

Project	COD$_{cr}$	BOD$_5$	NH$_3$–N	TN	TP	SS
Domestic sewage (mg/L)	400	200	40	50	5	200
Treated water (mg/L)	50	10	5 (8)	15	0.5	10

Note The value outside the brackets is the control index when the water temperature is greater than 12 °C while the value inside is the control index when the water temperature is smaller than or equal to 12 °C

5.11.2   Project Background

According to the "*13th Five-Year Plan for the Construction of Urban Sewage Treatment plants in Hainan*" issued by the Hainan Provincial Development and Reform Commission and the Provincial Department of Water Affairs in 2017, the comprehensive treatment capacity of urban sewage should be further improved to reach a centralized sewage treatment rate of more than 85% in the cities, counties and towns in Hainan province.

The project, located in Haikou, Hainan, involves the treatment of domestic sewage on a new campus. The sewage pipe network around the campus is not completed, and the domestic sewage generated on campus cannot be discharged to the sewage treatment plant. Hence, supporting sewage treatment plants are needed.

There are approximately 7000 faculties and students in the new school according to a field survey and data provided by school officials. The comprehensive domestic water consumption standard is determined as 200L/(person·d) with reference to the per capita comprehensive water consumption index of Haikou, Hainan, China, based on the current situation of school water consumption and long-term planning. And the design scale of the sewage treatment station on campus is 1500 m³/d based on the forecast of comprehensive water consumption and the construction of the sewage treatment station. Treated water shall meet the Level I A criteria of the *Discharge Standard of Pollutants for Municipal Sewage Treatment Plants* (GB 18918-2002).

5.11.3   Treatment Scheme

To meet the requirement of treated water quality, the A³/O-MBBR (mud-film composite) integrated sewage treatment process with satisfactory denitrification and phosphorus removal was selected. The process flow is shown in Fig. 5.19. The composition of the main structure of the project is shown in Table 5.21 Parameters of the integrated treatment device are shown in Table 5.22 The total sewage treatment capacity is 1500 m³/d, which is realized in the parallel operation of the A³/O-MBBR integrated sewage treatment equipment with a daily treatment capacity of 250 m³. The project was developed in two phases, and the case selected undertook

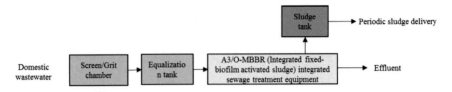

Fig. 5.19 Process flow of a domestic sewage treatment case on campus in Haikou, Hainan, Chi

the construction of the first phase with a design water volume of 750 m³/d, which involved three integrated facilities.

The main process flow is as follows: Domestic sewage collected by the pipeline flows into the equalization tank by gravity for homogeneous and equalized treatment after large suspended solids are removed by the screen. Then, the sewage is pumped

Table 5.21 List of main structures

Name	Dimensions (L × W × H)	Qty	Unit	Remarks
Equalization tank	12.0 * 5.2 * 6.0	1	Set	Built by the owner
Sludge tank	5.2 * 4.8 * 5.4	1	Set	Built by the owner
Equipment foundation	18.0 × 21.85 × 0.3 m	1	Set	Built by the owner
Discharge ditch	4.0 × 1.0 × 0.6 m	1	Set	Built by the owner

Table 5.22 Parameters for A³/O-MBBR process equipment

Project	Parameters	Remarks
Treated object	Typical domestic sewage	
Inlet of domestic sewage	Sewage inlet improved by the sewage submerged pump	
Rated capacity for sewage treatment (m³/d)	250	
Biochemical residence time (h)	8.32	
Air demand (m³/min)	4.0	
Type of carrier	Biomass increasing sponge	
Installed power (kW) air pump	4.63	
Operating power (kW) air pump	4.57	
Energy consumption per ton of water (kW·h/m³) air pump	0.34	
External material	SPA-H (Weathering steel)	White + gray
Dimension (L × W × H) (m)	17.5 × 2.9 × 2.9	
Net weight (t)	14.3	
Operating weight (t)	132.1	
Equipment installation mode	Ground type	

Fig. 5.20   Site of A domestic sewage treatment case on campus in Haikou, Hainan, China

into the Beisi efficient biological reactor. Treated water is sterilized and discharged after meeting the standards. And the excess sludge is discharged into the sludge thickening tank before transportation for landfill or composting. The integrated equipment for sewage treatment is installed on the ground in this project. The construction cost is 8500 yuan/m^3, with an operating cost of 0.7 yuan/m^3 (excluding the labor cost).

5.11.4   Treatment Effect

The A^3/O-MBBR process can achieve the project objective effectively. The phase 1 project with a treatment capacity of 750 m^3/d was put into operation, and the integrated equipment is stably operated with a satisfactory treatment effect. The quality of treated water can meet the Level I A Criteria of the *Pollutant Discharge Standard of Water for Municipal sewage treatment plants* (GB18918-2002). The site picture is shown in Fig. 5.20.

5.11.5   Tips for Operation and Maintenance

a. One person specializing in operation and maintenance should be arranged to regularly inspect the operation of equipment and process.
b. A daily water quality monitoring system should be established to regularly monitor the water quality of domestic sewage and treated water.

c. Sludge discharge of the equipment should be reasonably adjusted according to the water quality and quantity, so as to maintain the concentration stability of the sludge in the system.

d. Given the periodicity of domestic sewage discharge on campus, sludge discharge of all facilities is closed before the summer and winter holidays. Meanwhile, three air pumps of the parallel equipment are kept in intermittent operation during holidays to reduce the degradation of sludge in the lack of organic matters from domestic sewage, and industrial glucose should be regularly dosed in preparation for the smooth startup of equipment after the holidays.

e. The rest should refer to the operation and maintenance tips of the A^3/O-MBBR (Integrated fixed-biofilm activated sludge) process.

Printed in the United States
by Baker & Taylor Publisher Services